It was good enough for Buddha,
Plato, Socrates,
Hippocrates, Ovid,
Aristotle, Shelley,
Tolstoy,
George Bernard Shaw . . .

Here's news about food combining for better health . . . mythmaking and the protein-diet controversy . . . all about fasting . . . minerals, vitamins, supplements . . . meat vs. other proteins . . . and two food value tables precisely measuring the elements of nearly 500 common foods. . . .

Learn why vegetarian foods may be the best thing to happen to YOUR diet—and to your life!

PROTEIN FOR VEGETARIANS

GARY and STEVE NULL
and STAFF

REVISED AND ENLARGED

A JOVE BOOK

All tables in this book are from
the United States Bureau of Statistics

Seven previous printings
First Jove edition published August 1978

10 9 8 7 6 5 4 3

Printed in the United States of America

Jove books are published by Jove Publications, Inc., 200 Madison Avenue, New York, NY 10016

As with all books completed by Gary and Steve Null and the staff of the Nutrition Institute of America, every effort is taken to minimize contradictions and errors, and to present as definitive and complete a work as can be prepared for the subject matter being written.

The following staff members prepared the material for *Protein for Vegetarians*:

James Dawson
Chief Editor, N.I.A.

Richard Bull
Assistant Editor

Steve Null
Research Director

Diana Powers
Librarian for New York Public Library
Research Assistant

Jerry Krutman
Research Editor

Ollie Black, Rajah Afghan, and Edith Stern
Contributing Editors

As we increase our own awareness of nutrition and gather more new scientific research, we will update all of our books. Nutritional science is ever changing, and therefore what we know as a fact today could be a fallacy tomorrow.

The staff mentioned on each of our books includes many highly experienced and capable researchers. Those who assisted in spending months working on this book include: Ruth Snyder—M.A., Columbia University, who has spent twenty years in administration of nine school lunchrooms in public schools of Roslyn, N.Y., where 35,000 students are served daily. She is also an active member of National School Food Service Association, New York State School Food Service Association, Long Island Dietetic Association, New York State Teachers Association, Rosyln Administrators and Supervisors Association. Carolyn Lane, a registered nurse, has her M.A. degree and has authored several books. Other researchers include Pat Whitcomb, Lucille Pulitzer, Susan Burden, Connie Capacchione, and Bob Rocos.

ACKNOWLEDGMENTS

The following list of names is by no means complete. Rather, their research on the subjects of protein and vegetarianism was of particular value in the preparation of this book:

J. I. Rodale
Herbert M. Shelton
Bob Hoffman
R. F. Milton
F. Wokes
Hereward Carrington, Ph. D.
Barbara Davis
Frances M. Lappe
Ellen Buchmanewald
A. M. Altchul

For more detailed reading concerning the topics of food combining, fasting, and protein, we refer the reader to *The Complete Handbook of Nutrition* by Gary and Steve Null and Staff, published by Robert Speller and Sons, 10 East 23rd Street, New York, New York, 10010.

The tendency of writers on nutrition has been either to make their books so technical that the lay reader would find the book difficult to comprehend, or so general as to offer only a broad introduction. The purpose of each book in the series by Gary and Steve Null and Staff is to fill the gap between these two extremes by providing basic information to the lay reader in a step by step progression, gradually increasing the reader's knowledge of nutrition.

The *Complete Handbook of Nutrition* was the basic primer with capsulated observations on many of the more important areas of health and nutrition. We have followed this with separate books dealing in a comprehensive manner with important categories, i.e. *Food Combining Handbook, Body Pollution, Herbs for the 70's,* etc. *Protein for Vegetarians* is our latest effort in the ongoing series.

Due to our desire to have each book complement the best information of the preceding one, some of the information in every book is repeated from the preceding ones.

We feel that overlapping important facts will assist the reader in learning step by step, in a pyramidal way.

Gary and Steve Null
New York City
May, 1975

INTRODUCTION

At first glance, it may seem a bit incongruous for a doctor who has been recommending a diet with a considerable allowance for meat-eating in the management of obesity to be introducing a book for vegetarians.

But *Protein for Vegetarians* is by no means a book exclusively for vegetarians. Rather it is a book about how anyone might insure that he takes in an adequate amount and type of protein. Protein, in the long run, might be the most essential of nutrients. When nutritional problems on a world-wide basis are analyzed, the most difficult to overcome and the most prevalent deficiencies are usually those involving protein.

In our nation, the protein "shortage" has not been an overwhelming problem, but, because of economic factors and the world's food supply problems in general, it threatens to become so. When this begins to happen, as might first be evidenced by rising costs of meat, the reader who has absorbed the lessons contained within this book will be in the position of guaranteeing that he and his family will be much better equipped to provide for adequate intake of this vital portion of our food supply.

Protein for Vegetarians explains the currently accepted theories concerning the role which proteins play in the body's metabolic responses in a readable, informative fashion, but more importantly, it emphasizes the richness of the plant kingdom as an invaluable source of those needed proteins which we have tended to assume can only be found in the traditional animal sources. Vegetable proteins are not usually complete in providing all the body's amino acid requirements, but this book tells how they can be combined to provide a complete protein source. The book is particularly valuable to the non-vegetarian who merely wants to (or has to) cut down on his consumption of meat.

As always, the Nulls and their staff have researched their subject well, and the book contains a wealth of useful information. I would anticipate that its readers would know how to avoid protein malnutrition even in the presence of a protein famine.

Dr. Robert C. Atkins

TABLE OF CONTENTS

Chapter One

VEGETARIANISM

"More and more, in the next thirty years, we are going to be resorting to a vegetarian diet," Dr. Isaac Asimov predicted on a recent television talk show. The reasons for this, he explained, are not moral or humanitarian, but because plants remain our healthiest, largest, and most economical source of food. The noted writer and biochemist is not alone in making such a prediction.

A commission of the National Academy of Science in the United States has come to a similar conclusion. To survive in the future, the commission reported, we will all have to rely more on plant foods for sources of protein, instead of the traditional American diet of meat.

Ecologist Jarvis Eldrid of West Virginia University warned a meeting of nutritionists at Morgantown, West Virginia: "The day when man can wantonly prey upon the diminishing species with whom he shares this planet is nearly over. Let us hope the lengthening shadows are not specters of doom."

Vegetarianism is not new. It has been long established in India as an integral part of the Hindu religion. Plato and Socrates were vegetarians, as were Hippocrates, Ovid, Aristotle, and Buddha.

British poet and vegetarian Percy Shelley insisted in 1813 that "Man's digestive system was suited only to plant food." Fifty years later, British dramatist George Bernard Shaw wrote, "A man of my spiritual intensity does not eat corpses." On a lighter note, Russian novelist Leo Tolstoy said, "Feeding upon the carcass of a slain animal has something of a bad taste about it."

Vegetarianism has been misunderstood throughout most of history. Its practitioners have been called faddists, fanatics and mystics. There was seemingly something odd about a person who didn't eat meat. Hitler's vegetarianism, for example, has often been noted as one of his quirks. Even the word itself is widely misunderstood.

Etymologically, vegetarian does not come from 'vegetable,' but rather from the Latin word *vegetare*, which means "to enliven." When the Romans used the term *homo vegetus*, they referred to a vigorous person, sound in mind and body. Yet people today think of a vegetarian only as a vegetable eater!

The range of vegetarian diets is wide, almost as wide as the reasons people adhere to them. Some vegetarians eliminate meats from their diets, but continue to eat fish, poultry and dairy products; others eliminate meat and a combination of these foods. And many, of course, will eat only plant food. It would be unfair to say any one of these groups represents the 'true' vegetarian.

The reasons people give for being vegetarians are many. A few base it upon their religious beliefs, others claim health benefits. Most, however, came to vegetarianism out of rational judgment. It is this last reason which will primarily concern us in this book. We feel that vegetarianism is a rational alternative to the meat-centered American diet.

This book was not written for the orthodox vegetarian, but rather for any person who wishes to know how to use less meat in his diet. This new material is different from certain statements made in our other books, especially concerning fasting and food combinations. We feel that the areas of fasting and food combining are generally too confusing and regimented for the average person. Therefore, we have made certain corrections which will make eating more enjoyable and less restrictive.

The reader should not consider it a contradiction to have meat dishes in a book with a vegetarian theme. It is our interest here to show several alternative ways to consume protein. Also, understanding the eating habits of most people, it is best to gradually reduce meat from their diet.

We feel that too many popular nutritionists are too radical and are seeking overnight converts, rather than giving the public a systematic education about sound nutritional values and practices.

THE NEED FOR A RATIONAL ALTERNATIVE TO THE AMERICAN DIET

One of the strongest reasons for taking a serious look at our diet and seeking alternative menus is the proliferation of processed foods in this country over the past few years. The extensive processing of our foods causes many nutritional problems. While there is little chance a person living in America will starve to death, there does exist the danger of malnutrition.

Nutritionists, like the late Adelle Davis, are beginning to make us aware of the large gap between good health and the absence of illness. Most of us in this country fall into the gray area between these two categories. Unfortunately, Americans are *not* the world's best-fed people.

The title of a book published several years ago sums up two of the problems of the American diet: *Overfed and Undernourished*. Both overeating and undernourishment are detrimental to health, like overexertion and inactivity. In our case the two are paradoxically related.

We Americans consume large quantities of food but much of it is processed and gives us only naked calories. Therefore, the average person may have to eat three times as much food as he really needs to keep his system operating at peak performance. This excess causes many of us to suffer poor health.

Meat, of course, is one of the highest sources of proteins available, and Americans eat plenty of it. But few people realize that protein, vitamin B_{12}, phosphorus, potassium, and magnesium are practically its only nutrients. Plants provide more than twice the amount of vitamins and minerals as meat. Meat is high in both phosphorus and potassium, but many vegetables and dairy products are higher. Cheddar cheese is twice as high in phosphorus, and baked potatoes and lima beans are 30 to 50 percent higher. Also, the magnesium content of vegetables is much higher than meat.

With the exception of vitamin B_{12}, you can get all your vital nutrients from vegetable and fruit sources. B_{12} can be obtained from dairy products and eggs.

The nutrients of meat are not enough by themselves to keep you healthy and at peak efficiency. The same is true of other single foods. If you concentrate your diet on any one type of food, you'll suffer from a lack of certain elements. Only the foods containing a variety of essential elements are sufficient to sustain life. A vegetarian diet *will* provide all of these necessary elements.

Let's take a look at these elements. They include protein, fats, carbohydrates, mineral salts and vitamins, all of which humans obtain from food to replace worn-out body tissue.

The processes which keep us alive are those that wear out tissue. As the tissue is broken down the wastes are eliminated and the tissue itself replenished by the foods we eat. On the average the body remains at approximately the same weight. The complex nature of these tissues demands we replace the proper elements.

If only one essential element is missing from your diet, your health will suffer. This was dramatically illustrated in an experiment carried out by Harold Rossiter and Charlotte Beckner at Woodburn College in Pennsylvania. They put ten dogs on a diet primarily of fats, and all ten of them became round and plump. All died.

The absence of a single nutrient may cause pellagra, niacin deficiency, rickets, vitamin deficiency, scurvy, and beriberi. You need every nutrient to obtain and maintain optimum health.

Although protein cannot maintain our diets on its own, it is the most important element. It is probably also the most misunderstood. Many people mistakenly believe that a meat-centered diet is the only possible diet which supplies all the proteins and other nutrients your system needs. One rational alternative is to concentrate on getting all or most of your essential nutrients from plant sources.

Protein supplies most of the muscle-forming elements your body requires and acts as a fuel for energy. This second function can be as important as the first. About half of the protein you take into your body is transformed by the liver into the sugar, glucose. This process plays an important role in maintaining your energy level. You need fats and carbohydrates to supply heat and energy to your system and to influence protein metabolism. Vitamins and minerals are absolutely essential in keeping your metabolism working correctly.

How well your bodies are supplied with these essential elements and to what extent your systems use them have a profound effect on your life. Nutrition might not make you ambitious, but it can supply you with the energy to fulfill your ambitions. Poor nutrition can rob you of the energy you need to do the things you want. Recent research has shown that what you eat can even influence your mood and your psychological and emotional well being. *You Are What You Eat!*

REASONS FOR CUTTING DOWN MEAT CONSUMPTION

Can you get too much protein? There are many people who insist that you cannot. Recent evidence, however, shows that for some people a high protein diet might actually be harmful. This evidence falls into two categories.

Nutritionists argue that eating too much protein is a potential health hazard because of the toxic substances created by the presence of more protein in your system than you can use. Eating too many fats or carbohydrates can cause an excess that is not immediately dangerous. Proteins, on the other hand, immediately create uric acid and a variety of toxic substances which can poison and devitalize your system. An excess of carbohydrates will produce fermentation. An excess of protein results in putrefaction.

Another argument against a high protein diet is the discovery by Dr. George Watson, during research which lead to his book *Nutrition and Your Mind*, that some people have trouble oxidizing the substances taken into their bodies to produce energy. Proteins are oxidized at a slower rate than carbohydrates and fats, and if such a person has a heavy protein diet, he is compounding the problem and not getting enough energy to operate at a maximum level.

This brings up the very important point that the same diet is not necessarily adequate for everyone. A diet which meets all the needs of one person and allows him to operate at peak level of health may cause just the opposite effect in someone else. We've all heard of cases of individuals living to be very old on eccentric diets; but the reason for their reaching old age might not be the diet per se, but rather their particular metabolic makeup.

The old adage "One man's meat is another man's poison," is quite literally true. It is important that diet information be applied to oneself and patterned on one's own particular needs.

UNFIT FOR HUMAN CONSUMPTION

Meat is a complete protein, one of five different types of complete protein. Meat could be a good source of protein if it were not for the many toxic chemicals which are used so heavily by most meat producers. Aside from the high quality of protein, meat also contains decaying cell nuclei. The cell tissues of cattle contain several forms of naturally occurring poisons which are generally thrown off through the processes of elimination. However, it is impossible for the animal to throw off all of the accumulated toxins. If the animal were unable to eliminate the majority of these toxins it would die of self-poisoning. It is important to understand that along with the usable protein in meat you are also getting these biological waste by-products.

Toxins manufactured within the animal are not the only poisons which we're concerned with. Just as some of the pesticides you use end up stored in your bodies, and you have every right to be concerned about this, so do some of the pesticides taken in by animals get stored in their bodies and pass on to you when you eat their meat. A large portion of the poisons and pesticides you consume comes from the meat you eat. Taking a close look at the sources of pesticide residues in the United States diet, you'll see that you take in 0.281 parts per million of DDT, DDE, TDE pesticide residues through meat, fish, and poultry. You take in only 0.112 parts per million from dairy products. And in order of decreasing amounts: 0.041 ppm from oils, fats, and shortenings; 0.036 ppm from leafy vegetables; 0.027 ppm from fruits; 0.026 ppm from legumes; 0.008 ppm from grains and cereals; 0.007 ppm from root vegetables; and 0.003 ppm from potatoes.

There is a wide gap between the amount of pesticides you receive from meat, fish and fowl, and the amount taken in from the next lowest category, dairy products: a 0.169 ppm difference. In other words, meat, fish and fowl contain more than twice as many parts per million residue of pesticides than does the second highest category. The next highest sources of pesticide residues are oils, fats, and shortenings, which measure 0.041 ppm. Again, this is less than half of the next highest category.

By eliminating meat, fish and fowl from the diet and replacing these items with other forms of complete protein, you can reduce the amount of pesticide residue in your food supply by half. Of course you will have to consume twice the amount of plant protein to equal the reduction of animal protein in your diet.

In addition to pesticides, meat contains several other unhealthy additives. You would never sit down to a meal and order a plate of methoxychlor, chlordane, heptachlor, toxaphene, lindane, benzene, hexachloride, aldrin, dieldrin, with a side dish of DDT and garnished with pesticides. Or try an unappetizing plate of stilbestrol, aureomycin, mineral oil residue and a dozen other chemicals, all added to beef and stored in the fatty parts of meat. Regrettably, this is exactly what you would find in a piece of roast beef.

Of course most people enjoy their roast beef with gravy. Even the gravy will contain DDT and several other pesticides, in addition to antibiotics and toxins created by the interaction between the chlorine-dioxide bleach used on the flour and the flour nutrients.

THE CHEMICAL BARRAGE

It is a harrowing truth that today few animals in your meat supply escape the barrage of chemicals that, while they do not produce sudden death, poison your bodies. Can anyone insist that these chemicals are not likely to lead to chronic illness and slow deterioration of your bodies? Residues of these chemicals are passed on to every meat eating consumer. No other article in the American diet is as thoroughly tampered with as the meat you eat.

The average steer is conceived by artificial insemination. He is raised on an artificial sex-hormone implanted in his ear, fed with synthetic hormones, antibiotics, and insecticides, and shot with tranquilizers. He grazes on pastureland contaminated by radioactive fallout and pesticides. After spending his life on this chemical feast, he is then slaughtered for your table, offering the primary source of protein in the American diet. Not a very reassuring picture, is it?

The farm, where nature once had a balance, has become a factory geared to the chemical alteration of nature. No matter how disillusioning this might be to the romantics, the point is that these chemicals are being passed on to you in the food you eat. And it is unlikely that the picture is going to change drastically in the near future.

According to the June 1968 issue of *The Farm Journal*, still further chemical alteration might be in store for you. *The Journal* told the farmers that researchers are forcing activated charcoal down the throats of cows in order to trap some of the pesticide residues. The idea behind this is similar to using charcoal filters on cigarettes to trap the tars resulting from the burning of tobacco. This solution to pesticides in meat seems a little ridiculous, analogous to giving everybody a gas mask with activated charcoal filters to breathe polluted air instead of doing something about cleaning up the air in the first place.

The researchers do not stop there. These cows will also be fortified with treatments of the enzyme-stimulating drug, phenobarbital, which is available from any veterinarian by prescription.

Drugs are also used to slow down the cattle's metabolism so they'll put on more weight from less food. Up to 90 percent of all cattle going to market each year have received artificial hormones. But they are not alone. It is an accepted practice in the United States to feed antibiotics and tranquilizers to all our meat animals. One hundred million chickens are receiving these hormones. An antibiotic dip has been used to increase the shelf life of chickens and turkeys. And the Food and Drug Administration has recently given milk farmers permission to feed their cattle antibiotics.

It would be reassuring to know farmers were not allowed to add anything to beef cattle, chickens, or dairy cows which is hazardous to health. But this is simply not the case, as can be illustrated by the story of stilbestrol.

Stilbestrol is an artificial female sex hormone which farmers use for the simple economic reason that a few pennies worth of the hormone planted in a cow's ear results in an extra twelve dollars worth of beef. If this chemical is mixed with their feed, cattle gain weight 15 percent faster on 12 percent less feed. Stilbestrol is said to be worth 675,000,000 pounds of beef annually.

To date scientists can't explain exactly how weight is increased in beef or chickens with the female sex hormone stilbestrol. The Food and Drug Administration believes stilbestrol may affect the glandular system to cause the weight increase. In those animals where stilbestrol was used, "hypertrophy" (excessive development) of the liver, adrenal glands, and pituitary gland was noted.

There have been several outspoken critics of stilbestrol. One is Dr. Robert K. Enders, chairman of the department of Zoology at Swarthmore College. He is also an unsalaried advisor to the United States Department of Agriculture and the Department of the Interior. Dr. Enders admits that the use of stilbestrol makes the food more attractive but, referring specifically to its use in chickens, warns that "the use of the drug to fatten poultry is an economic fraud. Chicken feed is saved, but it is merely turned into fat instead of protein. . . . The fat is of very doubtful value and is in no way the dietary equal to the protein that the consumer thinks he is paying for." Not only has a chemical been added to your diet by using stilbestrol in chickens, but the chemical has lowered the quality of the product you buy. It raises the fat content and correspondingly lowers the protein content of the chicken. This would be bad enough, but there's more. Stilbestrol has been shown to be a health hazard. Among the many pathogenic conditions caused by stilbestrol are excessive menstrual bleeding, cysts and cancers of the uterus, arrested growth of children, fibroids of the uterus, breast pain in women, premenstrual tension, cancers of the cervix, breast and uterus and leukemia and tumors of the sexual glands in animals. Stilbestrol in men can cause impotency and sterility.

When stilbestrol was first tested in the early 50's on a group of volunteer women, no harmful results to their health were detected. Twenty years later the women were still in good health, but their daughters had developed vaginal cancer during adolescence. This test makes a strong case for those who question the long range effect of chemicals added to your food.

MORE CHEMICALS IN MEAT

The chemical feast forced on the livestock you eat does not stop at the farm. The list of chemicals commonly used in curing and preserving cold meats is a long one. And all these added chemicals are poisons.

Sodium nitrate and sodium nitrite are common in luncheon meats. They have both recently been shown to be cancer-causing agents. One of the reasons these chemicals are used is because they accentuate the natural color of the food, making it more appealing and more expensive. Both chemicals have been known to poison meat consumers.

One example of the inability to control the amounts of nitrites in meats was the several cases of illness in New Orleans, caused from frankfurters supposed to contain no more than 200 parts per million of nitrites. However, upon inspection the frankfurters actually contained up to 6,650 ppm of nitrites.

Before using any food which contains any amount of boric acid, you should know that the Food and Drug Administration considers it "poisonous per se". Yet there have been reported cases of boric acid being "dusted on hams during the curing process" to keep off a fly infection.

One of the most infamous of the surreptitious chemicals is sodium sulphite, a substance both dangerous and illegal. It is frequently used in manufacturing certain foodstuffs. The same is true of sodium benzoate.

John Cullen, a former Canadian food-inspection official, has warned,

> If the meat is of an unusually red color, it is reasonable to assume that it has been doped and doctored with sulphurous acid or sodium sulphite. This is especially true in the case of hamburger that has been made from stale meat trimmings, pork kidneys, pig's hearts, sheep hearts and other meat by-products including large quantities of fat. [These chemicals are

> a] great favorite with butchers and manufacturers of meat products generally. This preservative is very dangerous to health, especially when used in meat, because it will not only restore the color of putrid and almost black meat, but also because it will destroy the strong odor of putrefaction.

But this is not the end of Mr. Cullen's story. He goes on to say that

> many butchers will contend that they use this preparation only because it arrests the spread of bacteria. Nothing, however, is further from the truth. Changes of the most dangerous character are continuously taking place in the meat, but sodium sulphite obscures them and makes the meat appear to be fresh and of better value than it really is, and enables the seller to perpetrate an unscrupulous and deliberate fraud.

UNHEALTHY MEAT

Chemical additives can be very dangerous, but they are not the only consideration vegetarians take into account when explaining their case against meat as the mainstay of their diets. This is the hygienic argument we touched on earlier.

If a person were to watch cattle being slaughtered and prepared for sale, it would probably have some effect on his desire to eat meat. A little insight into the internal cell changes once the cattle are slaughtered should prove even more interesting.

Death is not immediate to a slaughtered animal. In fact, for several hours after the slaughter the tissues will be consuming the soluble food-material still in the bloodstream. The hygienic problems occur during these hours between the time the animal loses consciousness and the time all cell activity ceases. While the animal was alive those toxic by-products of elimination were quickly removed from the system through the lungs, liver, kidneys, and other excretory organs. Harm results when the waste causes the destruction of tissues and cells, after the somatic (bodily) death. The flesh of an animal carcass is loaded with toxic blood and other waste by-products. Cooking, aging or chemical additives can not extract from or lessen the effects of these poisons on the animal.

For the reader who is concerned about the extent of the adulteration of our total food supply, including meat, we refer you to *Body Pollution*, by Gary and Steve Null and Staff, published by Arco.

MEAT VERSUS OTHER SOURCES OF PROTEIN

The only important nutritional element we derive from meat is protein and its related substances. Meat lacks all other food factors essential to good health. There are other sources of protein that contain these other essential nutrients. Not only do they have a high percentage of protein, but in some cases they are a more economical supply of this important substance. Some of these foods are beans, peas, lentils, eggs, milk, cheese, whole grains, and especially nuts. The two highest protein nuts are cashews and almonds. Nuts are not only rich in protein, but also high in other important substances, mostly minerals and fatty acids.

One important argument against the place of meat in the national diet is the fact that much of the food which could be a direct source of protein to Americans is instead fed to cattle to produce meat protein and little else.

Livestock have the ability to convert inedible and low-quality material into high quality food for human consumption, but this potential is not exploited as much as it should be, especially in industrial countries like the United States. Enormous quantities of the *highest* quality food sources are fed to animals, instead. In this country one half of our harvested agricultural land is planted with feed crops. In fact, *we feed 78 percent of all our grain to animals.*

The high-quality food we feed to livestock is wasted. Cattle, sheep, and goats do not need to eat protein in order to produce protein. Micro-organisms in their stomachs can convert nitrogen in the form of urea into protein. The absurdity of feeding cattle very nutritious grains which humans could better use and afford can be seen by the De-

Table 1. Protein Supplies (1963-65) (per capita) and per day—by regions and subregions

Regions and Subregions	Calories	Animal Proteins	Vegetable Proteins	Total Proteins
FAR EAST (incl. China Mainland)	2,050	8.6	46.2	54.8
South Asia	2,020	6.4	43.0	49.4
Southern Asia Mainland	2,180	13.1	36.3	49.4
Eastern Asia	2,350	20.5	54.6	75.1
Southern Eastern Asia Major Islands	2,040	7.1	33.6	40.7
China Mainland	2,010	8.2	50.5	58.7
NEAR AND MIDDLE EAST	2,410	14.0	57.6	71.6
AFRICA	2,170	10.9	47.6	58.5
North Africa	2,100	10.9	44.1	55.0
West and Central Africa	2,120	7.8	46.9	54.7
East and Southern Africa	2,270	15.0	49.8	64.8
LATIN AMERICA	2,590	24.1	43.5	67.6
Brazil	2,780	19.4	49.4	68.8
Mexico and Central America	2,500	21.3	45.0	66.3
Northern and Western countries of South America	2,220	22.2	36.3	58.5
River Plate Countries	3,090	50.5	37.0	87.5
DEVELOPING REGIONS	2,140	10.7	46.9	57.6
EUROPE (incl. U.S.S.R.)	3,050	42.8	44.8	87.6
Eastern Europe	3,180	32.4	56.7	89.1
Western Europe	3,020	45.4	41.9	87.3
NORTH AMERICA	3,140	65.3	27.8	93.1
OCEANIA	3,230	63.9	31.5	95.4
DEVELOPED REGIONS	3,070	48.3	40.8	89.1
WORLD	2,380	21.0	45.1	66.1

Table 2. Percentage contribution of various commodities to percentage supplies (Protein supplies 1963-65)

	Cereals	Starchy roots and tubers	Pulses nuts and seeds	Vegetables and fruits	Vegetable proteins	Meat	Eggs	Fish	Milk	Animal proteins
FAR EAST (Incl. China Mainland)	59.8	3.3	18.0	3.3	84.3	6.6	0.7	4.6	3.8	15.7
South Asia	64.5	1.0	19.6	1.0	87.1	1.4	0.2	1.4	9.9	12.9
Southern Asia Mainland	58.8	2.0	8.3	4.0	73.5	7.1	1.4	15.4	2.6	26.5
Eastern Asia	48.2	2.1	14.0	8.4	72.7	6.1	2.9	15.6	2.7	27.3
South Eastern Asia Major Islands	64.4	6.4	7.4	3.9	82.6	7.1	1.0	8.6	0.7	17.4
China Mainland	57.8	4.6	20.3	3.2	86.1	10.0	0.5	2.7	0.5	13.9
NEAR AND MIDDLE EAST	67.8	1.0	6.7	4.9	80.1	8.0	0.7	1.4	9.5	19.6
AFRICA	54.7	9.1	15.7	1.9	81.4	9.2	0.5	4.1	4.8	18.6
North Africa	69.9	1.1	5.1	4.2	80.3	7.8	0.8	1.6	9.5	19.7
West and Central Africa	51.2	14.8	18.1	1.6	85.7	6.8	0.4	5.1	2.0	14.3
East and Southern Africa	55.1	4.5	15.6	1.7	76.9	12.5	0.6	3.4	6.6	23.1
LATIN AMERICA	39.8	4.0	16.9	3.4	64.3	18.3	1.9	2.7	12.7	35.7
Brazil	37.9	3.6	26.6	3.7	71.8	13.5	2.2	2.3	10.2	28.2
Mexico and Central America	44.3	2.1	18.2	3.0	67.9	12.7	1.8	2.4	14.9	32.1
Northern and Western Countries of South America	41.0	7.5	8.5	4.3	61.8	18.8	1.2	4.8	13.2	38.2
River Plate Countries	32.7	4.2	2.5	2.9	42.3	41.0	2.1	1.4	13.0	57.7
DEVELOPING REGIONS	57.2	3.8	16.8	3.3	81.4	8.3	0.9	4.0	5.4	18.6
EUROPE (Incl. U.S.S.R.)	36.8	5.5	3.8	5.4	51.5	21.5	3.8	4.2	18.8	48.5
Eastern Europe	50.0	6.4	3.0	4.2	63.6	16.4	2.5	1.5	15.1	36.4
Western Europe	33.5	5.4	3.9	5.6	48.4	22.8	4.1	4.9	19.8	51.6
NORTH AMERICA	17.6	2.6	4.6	5.2	30.1	36.3	5.8	2.9	24.9	69.9
OCEANIA	24.9	2.4	2.2	3.6	33.1	36.8	4.2	3.1	22.5	66.9
DEVELOPED REGIONS	31.9	4.7	3.9	5.3	45.8	25.4	4.3	3.9	20.4	54.2
WORLD	47.9	4.1	12.1	3.9	68.2	14.7	2.1	3.9	10.9	31.8

partment of Agriculture statistics for 1970. Grazing cattle were fed approximately twenty-three million tons of high quality protein, which produced only one million tons of beef protein. In effect, cattle raisers wasted twenty million tons of high quality grains which could have been used by humans.

LAND FOR MEAT OR LAND FOR PEOPLE?

In the United States it is hard to believe we'll ever have a shortage of food. The same is not true of the rest of the world, unfortunately, and might not be true, as Dr. Asimov pointed out, even in America by 2001.

The rising cost of meat seems to be like the weather: everybody complains, but nobody does anything about it. Just on the basis of how much plant protein goes into one pound of meat protein, the cost of one pound of meat protein is ten to twenty times higher than one pound of plant protein, without taking into account the added cost of manhours and equipment needed to get that one ton of meat protein.

If we take one acre of land to produce food, the choice of how we are going to use that land will be determined by several factors. One of them should be how much food we can get from that one acre by using it in different ways. If we're concerned with producing protein for our own consumption or for the consumption of a group, we have several choices. We can use the land for meat production, cattle grazing or feed growing for meat animals. However, if we use that acre for cereal production, we'll produce five times more protein for consumption. An acre planted with spinach will give us twenty-six times the amount of protein we would have if we devoted that acre to meat production.

Many authorities feel that we place too much emphasis on meat in our diets and should give more attention to the importance of vegetables, legumes, dairy products and grains. In fact, it has been estimated that 35–45 percent of the world's livestock is fed grains from which we could benefit nutritionally. If we conserved our grain supply and gave it to the poor and malnourished, instead of to cattle, we could easily feed nearly all of the chronically underfed people in the world.

Since we use between one-third and one-half of the continental land surface of the world for grazing and 40 percent of this land in producing vegetable sources suitable for human consumption, we can see that 15 to 20 percent of the continental land surface could be converted to producing food directly available for human use. The lands which produce vegetable sources of protein used for grazing could be turned into an inexpensive source of protein food if people relied less on meat and more on plant sources.

An unshaking emphasis on producing meat as the backbone of the American diet not only uses up vegetable sources of protein, but also other sources of food. In 1968 up to one-half of the world's fish catch was fed to livestock.

BUT ISN'T MEAT NECESSARY?

There are many people who consider plant protein inferior on two counts. First they assume meat to be the highest, richest, and most easily obtained source of protein. Second they believe that vegetable protein is inferior in quality and hardly worthy of human consumption.

Is meat actually the richest source of protein? Not necessarily. There are plants which rank higher than meat in the quantity of protein they provide, particularly in their processed form. *Soybean flour, for example, is over 40 percent protein, while meat is rarely more than 20 to 30 percent protein.* Some cheeses are also very high in protein. Parmesan cheese, for instance, is 36 percent protein. Meat, as a source of protein, ranks lower than these two sources, ranging from 20 to 30 percent protein, which is not much higher than dried beans, peas, and lentils. These three are essentially in the same range as meat, with 20 to 25 percent protein.

What about the quality of the protein you get from plants? This quality can be measured scientifically and computed into figures, called the Net Protein Utilization, or NPU. We will cover this term more closely in the next chapter.

The NPU lets you know what percentage of the protein you eat is actually available to your bodies, and is a good measure of the protein quality of your diets. The scale of foods in terms of Net Protein Utilization, based on a United Nations publication, ranges from 40 to 94.

The highest NPUs are 94 percent for eggs and 83–89 percent for milk. Although meat protein is represented for its quality and quantity of protein, it is not at the top of the list. Soybeans and brown rice and some cheeses are high in their NPUs.

In one aspect, animal proteins are superior to plant proteins because they more nearly match the requirements of the human body. This does not mean, however, that you must use meat as your main source of protein, because dairy products also contain these animal proteins. The only advantage of this protein is that you need to eat less meat than plants to fulfill your essential amino acid requirements.

But you have seen that plant sources of protein are many times more plentiful than meat sources on this planet, and that it is not necessary to rely fully on meat. You can get your supply of proteins from other sources. How this is done will be examined in the following chapters.

SUMMARY

There are several reasons behind the increasing popularity of meatless diets. Many of them are based on rational judgments rather than moral, religious, or faddist claims.

From the purely practical point of view, any area of land will supply more vegetable protein than meat protein. In many cases, the soil given to feeding cattle could supply twenty times more plant protein for human consumption if this land was devoted to grains and fruits and nuts.

It is possible to derive high quality sources of protein from plants. The protein deficiencies of one plant are easily covered by the richness of another. The amount of varied proteins you take in from plant sources can be sufficient to fill your needs.

By obtaining your proteins from a plant source instead of animal sources, you can reduce the amount of toxins taken into your body, which in turn should help reduce illness and possibly lengthen life.

There are some commonsense rules to follow when eating a vegetarian diet, but most of them apply to any diet:

Chew thoroughly, especially when eating nuts.

Do not eat when you are excessively tired, excited or worried. Such states can interfere with your digestive system.

Take a few minutes to rest before each meal so that your body can handle the food better. You should not exercise after eating.

Finally, do not think more food is required under a vegetarian diet.

A de-emphasis on meat can be a rational alternative to the American diet, if you are rational in applying it. The information in the following chapters should equip you with an understanding of your needs, especially for protein, and the facts about a vegetarian diet.

The next step is to take a close look at how your body handles proteins, how much protein you need and why, and what the best sources of proteins are. We will also discuss the combination in which foods can be eaten for proper digestion, assimilation, utilization, and elimination.

Chapter Two

PROTEINS, CARBOHYDRATES, AND FATS EXPLAINED: DIGESTION, ABSORPTION, AND METABOLISM

Since the dawn of time man has known instinctively that he must have food to survive. The search for edible foods was a major preoccupation of primitive man. Today, in an era of overstocked supermarket shelves, the search for nutritious, appetizing foods involves not only time but, more importantly, sound judgment.

Too few people give enough thought to the nutritional quality of their food purchases. Every human being must have adequate sources of nutrients, that is, protein, carbohydrate and fat, in order to grow or maintain itself. Food nutrients are chemical substance which supply nourishment. That fact cannot be said of many other chemically composed substances found in food today.

In order to compile a comprehensive book on the need and importance of protein in the diet, it is not only necessary to explain proteins in themselves, but also carbohydrates and fats. The proper balance of these nutrients in the daily diet cannot be stressed too strongly. One is not effectively utilized by the body without the others.

It is true that some books do not explain the relationship of a protein to a carbohydrate or a fat. However, such a book leaves the reader with an incomplete story about the body's food needs. This text presents a complete, comprehensive picture of essential nutritional needs. These needs become even more evident when formulated into a daily food guide by the United States Department of Agriculture.

A Daily Food Guide

Milk Group (8-ounce cups)
- 2 to 3 cups for children under 9 years
- 3 or more cups for children 9 to 12 years
- 4 cups or more for teen-agers
- 2 cups or more for adults
- 3 cups or more for pregnant women
- 4 cups or more for nursing mothers

Meat Group
- 2 or more servings. Count as one serving:
  - 2 to 3 ounces lean, cooked beef, veal, pork, lamb, poultry, fish—without bone
  - 2 eggs
  - 1 cup cooked dry beans, dry peas, lentils
  - 4 tablespoons peanut butter

Vegetable-Fruit Group (½ cup serving, or 1 piece fruit, etc.)
- 4 or more servings per day, including:
  - 1 serving of citrus fruit, or other fruit or vegetable as a good source of vitamin C, or 2 servings of a fair source
  - 1 serving, at least every other day, of a dark-green or deep-yellow vegetable for vitamin A
  - 2 or more servings of other vegetables and fruits, including potatoes

Bread-Cereals Group
- 4 or more servings daily (whole grain, enriched, or restored). Count as one serving:
  - 1 slice bread
  - 1 ounce ready-to-eat cereal
  - ½ to ¾ cup cooked cereal, corn meal, grits, macaroni, noodles, rice, or spaghetti

SOURCE: "*Consumers All* Yearbook of Agriculture, 1965," U.S. Department of Agriculture, Washington, D.C., 1965, p. 394.

Such a guide clearly shows that no single food can provide the body with every nutrient that is needed for optimum nutrition. Keeping that daily food guide in mind and the food groups that it outlines, can a person rate the nutrients in his diet? Are they adequate or inadequate?

The answer is yes. A person can and should be able to identify the nutrients in the foods he eats. Then and only then can a person be assured that he is getting an adequate combination of nutrients that are so essential for proper nutrition.

This type of nutritional awareness should not come from hearsay only. Too many people are struggling with misinformation in the wake of conflicting reports from the media about food. Nutritional awareness should grow first out of an understanding of how the body functions and, secondly, an understanding of the makeup and utilization of each food nutrient. Without this knowledge there is little motivation for proper nutrition.

By putting the imagination to work, it is fairly easy to understand how and why the body functions as it does. A person examining himself in a mirror sees the obvious, familiar features that compose a human body. If a person looks beyond the obvious he will see a totally new structure—a chemical structure.

This immense chemical structure that is referred to as a body constantly needs fresh chemicals in order to function. Essentially all matter is made up of chemical compounds. However, the body requires only a select group of chemicals to function. These chemicals are called nutrients. This chapter deals with three nutrients—protein, carbohydrate and fat.

In the supermarket, nutrients appear in a common form—food. Such as it is, food may be appealing to the senses, but the immense chemical being views it as something that is important only after it can be reduced to its smallest chemical form. The smallest chemical form of each nutrient is different, since nutrients are not necessarily composed of the same chemicals. Because the body needs all of the chemicals that compose nutrients, it must break every nutrient down to its unique, basic form before it can be used by the body.

The human body might even be called a sophisticated chemical laboratory. It accomplishes the task of reducing the nutrients to their smallest chemical form by digestion and absorption, thereby maintaining life. Digestion and absorption are merely preliminary steps for the most important body process related to food intake, that of metabolism. Metabolism refers to the individual ways the body uses the specific chemical compounds that result from the breakdown of food.

When the body breaks down nutrients to their simplest form, energy is released. This process is referred to as catabolism. The body needs this energy to perform the simplest tasks. The body also uses nutrients to build up body tissues and other substances. This process is anabolism. These two body processes are very much interrelated. Anabolism uses up energy to build up the body, but if the energy were not released by catabolism in the first place the body could not continually rebuild tissues.

In very general terms this explains why and how the

body uses food. Using this information as a basis for understanding, it is possible to speak in more specific terms about the relationship of three nutrients—proteins, carbohydrates and fats.

As far back as 1838 a chemical substance was described as the most important of all known substances, without which life did not appear possible. This complex substance was found to be present in food. It was called protein, from the Greek word meaning "to take first place." With the later development of the study of nutrition, proteins were analyzed more closely in the laboratory. Today, scientists are less inclined to maintain that protein is more important than any other nutrient. Nevertheless, the name protein was well applied.

There is now evidence that protein is a constituent of every living cell. Half the dry matter of an adult is protein. The only body constituents that normally contain no protein are urine and bile. An even more dramatic way to emphasize the importance of protein is to review the disheartening conditions that arise in countries without an adequate supply of protein foods.

As the population of the world expands, existing food supplies become more and more depleted. Literally millions of infants and young children are victims of protein-calorie malnutrition, a broad term that encompasses kwashiorkor and milder stages of malnutrition. Although many children do survive these diseases, they are unable to achieve their full physical development and may even suffer severe mental retardation.

Major government agencies, together with nutritionists and scientists, have turned their attention to the problem of exactly what quantity and quality of protein is necessary to sustain life and health. While an inadequate supply of protein foods does not appear to be a widespread problem at this time, it is a recognized fact that all the peoples of the world should be made more aware of the role of proteins in body functions, of daily protein requirements and of food sources to meet these needs.

When proteins were first discovered, it is unlikely that Mulder, the chemist responsible for isolating protein, had any conception of the extremely important roles this nutrient plays in the body, or even of the number and complexity of the protein components of the body. Proteins differ from carbohydrates and fats in that they contain nitrogen in addition to carbon, hydrogen and oxygen. Some proteins also have sulfur, phosphorus and iron. Proteins are unusual from a chemical standpoint because when they are broken down into a number of smaller structural units they may react as either a base or an acid. For this reason the smaller structures that make up protein were given a seemingly contradictory name, amino (base) acids.

A molecule of protein may contain as many as several hundred or even thousands of amino acids. These combinations of amino acids are connected together in a unique fashion known as the peptide linkage. For example, a given protein contains a specific number of certain amino acids, linked in a sequence that is specific for that protein. It is this very specificity of protein structure in a definite amino acid sequence that gives various tissues their unique form, function and character.

Proteins may be classified on the basis of their general structure and chemical nature. Simple proteins contain only amino acids or their derivatives. A few examples of this group are albumins and globulins, found in all body cells and in the blood serum; keratin, collagen and elastin, in supportive tissues of the body and in hair and nails; globin in hemoglobin; zein in corn; gliodin and gluten in wheat; lactalbumin in milk.

The second classification of proteins is the conjugated, or compound, proteins. These proteins are composed of simple proteins combined with a non-protein substance. Included in this group are lipoproteins, the vehicles for the transport of fat in the blood; nucleoproteins, found in large amounts in glandular tissue; phosphoproteins, such as casein in milk and ovovitellin in eggs; metalloproteins, such as the enzymes that contain mineral elements; chromoproteins, such as hemoglobin; mucoproteins, found in secretions from mucous membranes.

The final classification group of proteins are the derived proteins. Derived proteins are substances resulting from the decomposition of simple and conjugated proteins. They are labeled proteases, peptones and peptides.

The ultimate value of a food protein in the body lies in

its amino acid composition. Actually it is the amino acids that are the essential nutrients rather than the protein. To date approximately twenty-two amino acids have been discovered.

In laboratory experiments it has been shown that when specific amino acids are omitted from the diet, an animal subject will fail to grow or eventually die. Conversely, the elimination of other amino acids from the diet has no such harmful effects. Thus amino acids came to be classified as either essential or nonessential. The following chart lists the types of amino acids.

Amino Acids in Food and Body Tissue

CLASSIFICATION	AMINO ACID
NATURAL OCCURING AMINO ACIDS ESSENTIAL FOR ALL HUMAN BEINGS	Isoleucine
	Leucine
	Lysine
	Methionine
	Phenylalanine
	Threonine
	Trypotophan
	Valine
ESSENTIAL FOR INFANTS	Histidine
NONESSENTIAL	Glycine
	Glutamic acid
	Arginine
	Aspartic acid
	Proline
	Alanine
	Serine
	Tyrosine
	Cysteine
	Asparagine
	Glutamine
	Hydroxyproline
	Citrulline

An essential amino acid is one that cannot be manufactured or synthesized by the body at a rate sufficient to meet the needs for growth and maintenance. On the basis of such body dependence, eight amino acids have been demonstrated to be essential for adults. As shown on the chart, a specific amino acid is also thought to be essential for infants. The remaining amino acids are nonessential. The body can manufacture them and they are not as necessary for normal growth and development.

Many foods commonly designated as proteins are more accurately called protein-rich foods. Their nutritive value lies in the amino acid composition of various proteins. For this reason proteins in food are classified also on the basis of their amino acid content. Although there is some overlap in the classification, this grouping of proteins provides a simple basis on which to judge protein quality.

Proteins with enough of the essential amino acids to maintain body tissue and promote growth are called complete proteins. In terms of quality, these are the most desirable types of protein that can be supplied in the diet. Another type of protein may contain all the essential amino acids but a relatively small amount of one. This type of protein is said to be partially complete. It can maintain life, but it lacks sufficient amounts of some of the amino acids necessary for growth. A complete protein may at some times function as a partially complete protein if it constitutes only a small portion of the diet.

Incomplete proteins are low in quality. They are labeled such because they lack one or more essential amino acids. When this type of food is the sole source of protein in the diet, no new tissue can be formed, nor can worn out tissue be replaced. A very few foods are classed as totally incomplete proteins. They can neither build new tissue nor promote growth. A classic example of such a protein is gelatin.

The former classifications of proteins makes it quite obvious that certain types of proteins are more desirable than others. If proteins are to function in the body not only to maintain life but to promote growth, they must be supplied from good quality sources. All animal proteins, except gelatin, are complete proteins and, therefore, should be eaten in sufficient quantity to supply the body with all the essential amino acids.

Wheat germ and dried yeast have a biological value approaching that of animal sources. All other vegetable proteins, except nuts, are incomplete proteins. As such, these foods alone cannot supply the body's total need for protein. The poor quality of an incomplete protein can be offset if enough complete proteins are also included in the diet.

Although we depend on protein-rich foods as a source of amino acids, it is the specific type and amount of amino acids provided that determines how effectively a body reacts to use the protein available. The need for amino acids is evident in the numerous ways the body functions with the help of protein.

Perhaps the most important function proteins perform is promoting growth and body maintenance. Proteins supply the necessary amino acid building blocks used to create new tissue and repair worn out tissue. No other nutrients can do this because the essential building blocks of tissue are available only from protein. This is the reason for an increased protein need during periods of rapid growth, as in infancy, childhood and pregnancy.

Body proteins are necessary for the formation of essential body compounds. Enzymes are proteins. Some hormones are protein in nature. Mucus and milk are largely protein. Sperm is largely protein, as is the fluid in which sperm is contained. The antibodies of the body, responsible for the ability to combat infection, are protein substances.

Hemoglobin, the substance that performs a vital role in carrying oxygen to the tissues, is a protein complex. The plasma proteins help regulate the distribution of fluids both inside and outside of cells. Since proteins are capable of reacting with either acids or bases, their presence in the blood is vital to prevent the accumulation of too much acid or base, either of which would interfere with normal body functioning.

It is also possible for proteins to function as a source of heat and energy for the body. They supply four calories per gram of protein, the same as do carbohydrates. Under normal conditions, when an adequate amount of each nutrient is supplied in the diet, the constant breakdown and replacement of amino acids result in 58 percent of the protein intake becoming available for energy, hence four calories per gram.

However, it should be remembered that proteins have more important tasks to perform than that of supplying the body with energy. When adequate amounts of carbohydrates are provided for energy, proteins can then participate in body functions specific only for protein. This relationship between proteins and carbohydrates has caused carbohydrates to be referred to as protein-sparing foods.

Few people will say no to a succulent steak dinner. They wholeheartedly enjoy the sight, the smell and the taste of good meat as it fills up their senses. At the end of such a meal, ideally, there is nothing better than settling back and idling away an hour or two. Throughout this period of time when a person may be leisurely enjoying himself, the body is far from idle. It is performing the very meaningful task of digestion.

After a protein-rich meal the digestive process sets to work to convert proteins to amino acids so that they can be absorbed. Besides being concerned with the food just eaten, the digestive process must also act on a sizable amount of proteins that are constantly being released from the worn out cells of the mucosa.

The process of digestion of proteins occurs in two steps. In the mouth, foods are broken up by chewing, mixed with salivary secretions and passed as a semi-solid mass into the stomach. This is called mechanical digestion. Chemical digestion, the next process that takes place, begins in the stomach. Three enzymes contained in the gastric secretions participate in different ways in this beginning breakdown of protein. A given enzyme is capable of splitting only specific peptide linkages. It does not attack each and every linkage.

An inactive substance, pepsinogen, is produced by a single layer of cells in the mucosa of the stomach wall. Gastric hydrochloric acid acts as a catalyst to convert pepsinogen to the active enzyme pepsin. Pepsin is the main gastric enzyme specific for proteins. The active pepsin begins breaking the peptide linkages of protein to produce proteoses and peptones. These are the shorter chain of polypeptides which are still rather large protein derivatives. The action of pepsin completes the beginning stage of chemical digestion.

The next stage of chemical digestion takes place in the alkaline medium of the small intestine, with a number of enzymes from the pancreatic and intestinal secretions taking part. Protein-specific enzymes act on the already shortened chain of polypeptides at specific points until simpler, short-chain peptides are produced and free amino acids are split off from the ends of these chains. Amino acids are then ready for absorption by the intestinal mucosa.

The end products of protein digestion are water soluble amino acids. These amino acids are in sufficiently simple form chemically to pass from the wall of the intestinal tract into the bloodstream. They are carried by the portal vein to the liver. Absorption takes place either by diffusion or by the energy-requiring process of active transport.

It is interesting to note that competition for absorption seems to exist among amino acids. Absorption occurs at different rates, the premise being that a slow process will provide for a more uniform and efficient rate of release. When a mixture of amino acids is fed to an organism, the quantitatively predominant amino acid may retard the absorption of the others.

In the laboratory, it has been demonstrated that the amino acids methionine, leucine and isoleucine have the highest rates of absorption. Therefore, it is suspected that a protein that might contain an excess of leucine would be less effective than a more balanced protein. A point of importance that cannot be forgotten in dietary planning is that some protein of good quality should be present in each meal so that the available mixture of amino acids at any given time is optimum in quality as well as quantity.

The process by which the body will use the digested and absorbed amino acids is referred to as metabolism. There are actually many fascinating metabolic processes involved in protein metabolism. They can be explained and understood more clearly when separated into three categories—balance, building tissue and breakdown of tissue.

The balance category refers to the unique chemical composition of protein. Proteins are the only nutrients containing nitrogen. Thus studies of nitrogen metabolism have become the basis for assessing protein metabolism. Nitrogen balance studies are based on the fact that protein, on the average, contains 16 percent nitrogen. The nitrogen content is determined chemically and this figure, multiplied by 6.25, gives the amount of protein present in the substance. If the amount of nitrogen that goes into the body in food and the amount that leaves the body in the excreta are determined, what has been used by the body can be calculated.

Different states of nitrogen balance can occur in the body. If the nitrogen intake and the nitrogen output are equal, an individual is in nitrogen balance. This state is known as dynamic equilibrium. Body proteins are not rigid structures. There is a continuous rate of taking up and release of amino acids, an equilibrium. The liver, being the key organ of protein metabolism, has a high rate of amino acid turnover. The liver uses amino acids to build up its own protein compounds such as lipoproteins, plasma albumins, globulins and fibrinogen.

Muscle proteins have a much slower rate of turnover, and that of the brain cells is negligible. However, with the aid of proteins the intestinal mucosa renews itself every one to three days. This constant flow of amino acids being transported in the circulation to the cells and from the cells back into the circulation results in an amino acid pool, whereby there is a supply of amino acids to the tissue at any given location, at any given moment. Those amino acids available from dietary sources are exogenous proteins. Those coming from tissue breakdown are endogenous proteins.

Two other states of nitrogen balance can occur besides equilibrium or nitrogen balance. Positive nitrogen balance occurs when the intake of nitrogen exceeds the excretion. It exists during periods of growth, in infancy, childhood, adolescence, during pregnancy when the fetus is growing or when a nursing mother is storing proteins for breast feeding. Positive nitrogen balance also occurs during illness or injury or when muscles are being developed during athletic training.

When more nitrogen is excreted than is taken in, a state of negative nitrogen balance exists. This condition is not at all desirable. A person is losing nitrogen more rapidly than it can be replaced. A negative nitrogen balance can be brought on by fever, surgery, toxemia, burns, or shock. It may also occur if the calorie content of the diet is inadequate and tissue protein is being broken down to supply energy. A diet of poor quality protein that cannot meet the body's needs for tissue replacement, may also lead to a negative nitrogen balance.

When negative nitrogen balance occurs, the body can maintain itself for a substantial length of time. Although it would be inaccurate to say that the body stores protein to meet emergencies, it is true that certain reserves are available from practically all body tissues. Vital body functions may be protected for up to thirty to fifty days of total starvation or for much longer periods of partial starvation. However, these reserves eventually require restoration.

The major value of determining the nitrogen balance of an individual is that this knowledge can be used to determine the body's needs for proteins. Recommended dietary allowances, which will be explained in detail at a later point, are based on a consideration of the nitrogen balance of the body.

The two remaining catagories that better explain the metabolism of proteins are anabolism and catabolism. There are several determining factors that influence whether a protein will be used for tissue buildup or breakdown. These factors are a direct result of the total physical state of an individual and the amount and quality of amino acids present.

The body responds to an increased need of tissue buildup during growth and convalescence. In the adult this buildup of tissue just balances tissue depletion when the calorie intake is adequate. If the caloric intake is not sufficient to provide for energy needs, protein synthesis will not proceed at an optimum rate.

Even when the calorie intake is adequate, proteins may be unable to be used for tissue buildup, because the body is governed by the "all or none" law. This law requires that all the amino acids for the synthesis of a specific protein must be present in the amino acid pool at the same time. If the supply of one vital amino acid is limited, protein can be formed only so long as the supply lasts. When the supply is depleted the remaining amino acids will be used only for energy.

At certain times the need for specific tissue development may override the fact that the body may be in a state of negative nitrogen balance and proteins continue to be used for tissue building. This can be seen in some pregnancies when the fetus of an unborn baby depletes the maternal protein supply for its own tissue building, if the mother's diet is inadequate. Rapidly growing tumors will also use amino acids for malignant tissue growth instead of normal tissue buildup.

Protein catabolism is greater immediately after injury or burns and is also increased as a result of stress. At different times hormones may either stimulate the breakdown of protein tissue or enhance tissue buildup.

Tissue building proteins are in great demand by the various cells. Each cell requires specific proteins for its own function. The specific protein that will be used to synthesize or build up a compound is governed by the genetic code. This set programming exists because of deoxyribonucleic acid (DNA) that is found in cells. DNA has been the subject of dramatic research studies by scientists in their search to unravel the life processes of body cells.

The DNA genetic code exists at the center of a cell. However, the synthesis of a specific protein program takes place in the cytoplasm that surrounds that core. The plan for protein synthesis must be carried from the DNA molecules to the cytoplasm before a protein can be programmed. This is accomplished by the messenger ribonucleic acid (RNA).

Amino acids enter the cell and are activated by an activating enzyme which is specific for each protein. They join with another type of RNA, known as transfer RNA, which is also specific for each amino acid. The RNA transfer complex moved to the messenger RNA where the new peptide linkage is formed, according to the pattern carried from DNA. At this point, a protein is free to perform its function. The daily synthesis of protein in the adult is estimated to be about 1.3 gm. per kilo gram or about 91 gm. for a 154 lb. man.

If a given amino acid is not used in tissue protein synthesis it is returned to the liver. In the liver a process called deamination takes place which separates the amino acids into a nitrogenous residue and a non-nitrogenous residue. The nitrogen portion undergoes a series of chemical changes and is converted into urea by the liver and excreted in the urine.

The non-nitrogen residue is referred to as keto acids. Keto acids are ultimately used by the body for energy, but before that happens the keto acids enter the common metabolic pool where they may be synthesized for carbohydrates or fats. This action has resulted in the terms glycogenic (leading to the formation of a carbohydrate) and ketogenic (leading to the formation of a fat) being applied to some protein derivatives. Thus, even catabolism further illustrates the body's constant metabolic interconversions between protein, carbohydrate and fat.

Since protein is such an essential nutrient, much study has been given to the question of how much protein the body actually requires. Historically, wide variations in protein recommendations have been made, much to the confusion of the general public. With the recognition that the body needs amino acids and nitrogen rather than protein as such for protein synthesis, the term "protein requirement" is becoming outmoded. Nitrogen balance studies have been used in an attempt to determine exact, daily essential amino acid needs. However, there is still much controversy over proposed values.

A widely used index is the biological value (BV). This term, which refers to nitrogen balance, or more specifically nitrogen retained by the body, is often heard in discussions about protein nutrition. It has much the same meaning as incomplete and complete proteins in that it is taken to mean the quality based on specific amino acids.

The biological protein of egg is usually close to 100, so it was chosen as a standard or interpreted as having the highest biological value. A protein with a biological value of 70 or more is considered capable of supporting growth, assuming that the calorie value of the diet is adequate. This index may be applied to single proteins, single foods or combinations of proteins in food. The BV is based on the amount of nitrogen absorbed and does not take into account differences in digestibility from one protein to another.

Another guideline using biological value as a reference point is the provisional amino acid pattern outlined by the Food and Agricultural Organization, a division of the World Health Organization under the United Nations. The most interesting value that comes to light in this guide is that, again, whole egg and human milk have amino acid patterns that correspond most closely with the pattern required by humans. This pattern was based on the kind and amount of amino acids required by a healthy person to meet obligatory nitrogen losses in urine, feces and skin. The reference protein used has a biologic value of 100 and produces 1 gm. of tissue for each gram consumed.

The measurement of protein quality by an index known as net protein utilization (NPU) has been introduced to express, in a single measurement, both the digestibility of the protein and the biological value of the amino acid mixture absorbed from the intestine.

The PEM or protein efficiency method for determining protein quality is based on the assumption that weight gain of a growing animal is in proportion to gain in body protein. The fact that so many standards for evaluating the amino acid quality of protein have been proposed is evidence that no single, satisfactory standard exists at present. In practical nutrition, the focus is still very much on the consideration of daily protein requirements.

A source of confusion in the determination of requirements is the unclear terminology applied to standards. General terms such as minimum, average, adequate and optimum can be misleading until there is a sound basis for understanding. To facilitate understanding, the following discussion will confine itself to "recommended allowances." The Food and Nutrition Board of the National Research Council's published standards for protein is the case in point.

The protein allowance for adults recommended by the Food and Nutrition Board is 65 gm. for the 153 lb. man (70 kg.) and 55 gm. for the 58 kg. woman. That comes to 9 gm. per kg. of body weight. For persons who wish to figure personal protein allowances, it is recommended that protein allowances be based upon the ideal weight, that is, what a person should weigh, not what he does weigh.

Protein allowances were determined on the basis of nitrogen balance studies. The body's need for protein remains fairly constant, yet it can vary as a result of certain stresses or demands. To compensate for the needs of people who have a higher than average requirement, the Food and Nutrition Board calculated that the average, minimum requirement, a low 35 gm. plus 30 percent, would meet the needs of most individuals. This increased the protein allowance for a man to 45.5 gm.

After further thought, the Food and Nutrition Board calculated that the efficiency of a protein might be limited in many diets. To provide for such a situation, the allowance of 45.5 gm. was raised to 65 gm. Only after these steps had been taken was the figure 65 gm. released as the "recommended protein allowance." The protein allowances for various ages, both sexes and for pregnancy and lactation are indicated in the following chart.

Recommended Daily Dietary Allowances for Protein, Revised 1968

Designed for the maintenance of good nutrition of practically all people in the U.S.A. (Allowances are intended for persons normally active in a temperate climate.)

	Age (years)	Weight (kg.)	Weight (lbs.)	Height (cm.)	Height (in.)	Kilocalories	Protein (g.)
Infants	0–1/6	4	9	55	22	kg. × 120	kg. × 2.2
	1/6–½	7	15	63	25	kg. × 110	kg. × 2.0
	½–1	9	20	72	28	kg. × 100	kg. × 1.8
Children	1–2	12	26	81	32	1100	25
	2–3	14	31	91	36	1250	25
	3–4	16	35	100	39	1400	30
	4–6	19	42	110	43	1600	30
	6–8	23	51	121	48	2000	35
	8–10	28	62	131	52	2200	40
Males	10–12	35	77	140	55	2500	45
	12–14	43	95	151	59	2700	50
	14–18	59	130	170	67	3000	60
	18–22	67	147	175	69	2800	60
	22–35	70	154	175	69	2800	65
	35–55	70	154	173	68	2600	65
	55–75+	70	154	171	67	2400	65
Females	10–12	35	77	142	56	2250	50
	12–14	44	97	154	61	2300	50
	14–16	52	114	157	62	2400	55
	16–18	54	119	160	63	2300	55
	18–22	58	128	163	64	2000	55
	22–35	58	128	163	64	2000	55
	35–55	58	128	160	63	1850	55
	55–75+	58	128	157	62	1700	75
Pregnancy						+200	65
Lactation						+1000	75

SOURCE: *Recommended Dietary Allowances*, 7th ed., Food and Nutrition Board, National Academy of Sciences-National Research Council, Washington, D.C., 1968.

These so called protein allowances are not as rigid as they may appear. The National Research Councils' published standards are intended to serve only as recommendations. In order to cover a wide range of need and in order to provide a margin of safety to cover stress situations, they are approximately double the minimum protein requirements. As such, they should serve only as rough guidelines, to be adapted to a person's individual needs.

Of necessity, people in many countries in the world exist on protein intakes below that of 9 gm. per day, per kg. of body weight. An adult can, if necessary, get along on protein intakes as low as 30 to 40 gm. daily, depending on the quality ingested. Coincidentally, on such a low protein intake the urinary nitrogen output falls drastically, which indicates an adaptation process to compensate for the low protein intake. Equilibrium reestablishes itself at a lower level, unless the decreased intake of protein is below the critical point.

Recommended protein allowances have little meaning unless they can be put to practical use. Such practical application comes into play at mealtime. What foods should be provided to supply the best sources of protein? The animal sources such as milk, eggs, cheese, meat, poultry and fish have high biological value. Three of these foods alone can provide 51 gm. of protein. They are 1 pint of milk, 1 egg, and 4 ounces of meat or meat substitutes (fish, poultry, cheese). Grains such as wheat, soybeans, corn, rice, as well as nuts, peas, beans, lentils and peanuts provide a second source of protein.

Since there is a wide biological value among specific food sources of proteins, the obvious way to provide for the maximum, effective use of protein in the diet is to eat a combination of protein foods at each meal. Many popular food habits reflect this application of supplementing food proteins. Macaroni and cheese, casserole dishes containing legumes with small amounts of meats, sandwiches that combine grain in bread with meat, vegetables or cheese, and cereal with the addition of milk are satisfactory ways of balancing the quality of proteins.

As a rule, when appreciable amounts of plant proteins are eaten with a small amount of animal protein, the quality of the mixture is as effective as if only an animal protein had been eaten. Certain plant foods such as corn and dry beans, when eaten together, significantly improve the biologic value of proteins. This is because all plant foods are not deficient in the same amino acids and one may provide what another food lacks. Pure vegetarians can, therefore, obtain a satisfactory combination of amino acids from a diet that consists of whole grains, nuts, legumes and vegetables.

Although the protein intake of many people is adequate in quality, the spacing of these foods must not be ignored. It should be remembered that a good quality protein must be available at all times in the body if optimum protein utilization is to occur. A good rule to follow is to provide a complete protein at each meal. It is especially important to have a complete protein food at breakfast. Because breakfast is the first meal of the day after a long fast, a person needs to eat a protein of high biologic value at that time.

Protein foods are by and large the most expensive items in the diet. However, it is possible to economize and not sacrifice quality. The cut of meat used will not effect the nutritional value. It is possible to eat 4 oz. of stew meat and obtain the same value protein as is found in 4 oz. of steak. Also, the meats of all animals, poultry, beef, lamb, pork, any organ meats and even fish contain approximately the same amount of protein.

Legumes have often been called the poor man's meat, because when combined with small amounts of animal protein they add good protein value to the diet and good dollar value to the budget. It is also possible to economize in the selection of milk. Dry skim milk is as high a nutritional value as whole, fresh milk in terms of protein content.

A note of caution must be stated about the proper cooking of protein foods used in the daily diet. Research has shown that overheating or overcooking with dry heat impairs the nutritional quality of proteins. Marked losses of amino acids and decreased solubility and digestibility are attributed to the effects of dry heat on a variety of proteins. The biological value of proteins is best maintained when a meat is moderately cooked. Moderately cooked meat is also better digested than raw or overcooked meat. Cooking in water or steaming increases the digestibility of some proteins and may even liberate certain amino acids. A good example of this is the fact that soybeans and wheat, in particular, have a higher biological value when cooked.

Not only is proper cooking of proteins essential in order

to obtain the best biologic value, in some cases it may be a question of reducing fatal toxic substances. Nature has seen fit to contaminate foods with chemicals that are often toxic to man. Unlike chemicals which are deliberately added to foods, these natural toxicants pose a special problem since their presence cannot be legislated out of existence.

In terms of human nutrition toxic substances have, in most cases, been found to be controllable to the extent that heat reduces most toxins to a state where they are considered harmless. Soybeans, navy beans, lima beans and the peanut contain naturally existing toxins. Preliminary soaking prior to cooking and thorough cooking is necessary in order to destroy these potentially dangerous toxins. The use of bean flours for baked goods should also be viewed with caution.

Some protein foods normally considered to be nontoxic and, in fact, quite nutritious may prove to be harmful in special circumstances. Cheese contains large amounts of amino acid derivatives that result from the fermenting action of bacteria. These amino acid derivatives are toxic in nature, but considered harmless because the body normally detoxifies these substances with no trouble. However, some antidepressant drugs inhibit this detoxification process. Deaths have been reported in people who ate even small amounts of cheese while using certain antidepressant drugs.

Nitrites and nitrates used in the curing of meats and as a preservative can combine with harmless, naturally occurring compounds in meats to produce a substance that is known to be a potent liver carcinogen. Bacon, ham, salami and sausage, when preserved with nitrate and nitrite, contain small levels of this toxic substance. To date these small levels have not been proven lethal. More studies are needed about the cumulative effects of such hazardous foodstuffs.

Experimental evidence may be produced to indicate that almost any food substance may be harmful under certain conditions. It is little wonder that people need to be educated to maintain constant vigilance in both the supermarket and the kitchen.

There is a constant need for information about food

products in the American diet. On a broader scale, more information is needed by underdeveloped countries. Food sources of good protein are very scarce in many parts of the world.

Meat is the main source of protein only in the United States, Argentina, Australia, Canada and New Zealand. About 70 percent of the world's protein is estimated to come from vegetable sources. In a number of instances, efforts have been made to develop sources of proteins to enrich poor quality diets.

Fish meal has for a long time been used as a high quality animal protein in various animal feeds. Successful efforts have been made by the United Nations to mix fish protein with various foods in the diet of people in countries where protein malnutrition is a problem.

The nutritional value of cereal proteins is limited by a deficiency of one or more amino acids. Lysine is the first limiting amino acid in wheat. In fish protein there is a surplus of lysine. By adding increasing amounts of fish protein to wheat flour, its nutritional value is enhanced.

Fish protein has been incorporated, with the greatest success, into various cereal products such as bread, cookies, and pasta. Since protein malnutrition in developing countries is most prevalent in the 6 months–3-year-age group, weaning foods containing fish protein were also produced, and used in India and Bangladesh. Other countries have used fish proteins in rice, soup and chocolate beverages.

Investigation of the United States market for protein additives has shown that there is the likelihood that fish protein could eventually compete in the U.S. market as a protein ingredient in beverages, breakfast foods, canned and processed meats and baked goods. However, the use of currently available fish protein concentrates is usually confined to cereals, breads and pasta because of its limiting characteristics. Fish protein is a grayish powder with no texture, low wettability and low fat absorption. If more functional properties could be developed, along with an acceptable taste, fish protein could be used where nonfat dried milk is now used. This depends on further technical developments.

The future of the use of fish protein concentrates for

their nutritional properties is very dependent on the decisions made by the government authorities involved in aid programs. Studies have reinforced and increased worldwide awareness about the protein gap in the world food problem. Money and manpower are now needed so that developing countries can utilize science and technology to meet their food needs. At local levels there is a constant need for a program to improve nutritional knowledge.

While many Americans are sitting back, secure in the knowledge that they live in the best-fed country in the world, public health agencies are continually uncovering evidence that shows the lack of understanding people have about the importance of protein. Statistics very clearly show that protein deficiencies do exist in many vulnerable groups in the United States.

The high rates of anemia, premature births, and miscarriages among women of low economic groups may be directly related to protein deficiencies. The elderly who, because of low incomes or lack of incentive to eat, and the chronically ill who have little appetite are also likely to have insufficient protein in their diets. Although they may not display the obvious physical symptoms of malnutrition, these people very often have prolonged and difficult periods of convalescence following infections or surgery.

Simply because the protein intake in the United States is high is no reason to become complacent about good nutrition. In fact, everyone is vulnerable to protein deficiencies in the sense that, without proper education and understanding of nutrition, it is very easy to fall into poor eating habits in the exact situations mentioned. The future may demand that everyone have a sound knowledge of good nutrition to fall back on in times of uncertain economy and uncertain food supplies. Before any problems become a reality it is time to put into practice the essentials of good nutrition.

CARBOHYDRATES

Whereas protein foods have been shown to be of first importance, carbohydrate foods might well be called the fuel of life. Carbohydrates provide the energy which is needed to perform daily tasks. These foods are the starches and sugars.

People the world over obtain their major source of calories from carbohydrates. The chief sources of carbohydrates are grains, vegetables, fruits, syrups and sugars. Because these foods are often inexpensive and easy to obtain the amount of carbohydrates consumed tends to be greater at the lower economic level.

By examining the natural sources and the forms of the most common carbohydrate foods used around the world, a picture emerges about the respective standards of living in various countries. In Asia, the predominating carbohydrate is rice; in Europe, wheat and occasionally rye. Potatoes are important in Poland, Scandinavia and Russia. Central and South America rely heavily on corn for their carbohydrate food.

Surprisingly enough, in the United States the amount of carbohydrates in the daily diet has decreased by comparison with sixty years ago. More interesting yet is the type of carbohydrates that is being less often consumed. The amount of cereals, breads and potatoes in the diet has substantially decreased, while the use of sugars and sweets has gradually increased. This change in eating habits reflects not only the effects of technology, economy and the marketing system but also the inadequate nutritional status of many individuals.

Carbohydrates in their simplest form are composed of the three elements of carbon, hydrogen and oxygen. Plants store carbohydrates as their chief source of energy. They do this by the complex process of photosynthesis.

The chlorophyll in leaves, which gives plants their green color, actually harnesses some of the sun's energy so that it can react with the carbon dioxide of the air and the water of the soil to form carbohydrates which are stored in the plant and in turn used by animals. This is nature's first step in the manufacture of all foods. Man benefits from the simple sugars or starches found in grains, vegetables and other plant forms because these carbohydrate nutrients can be converted into fuel or energy by the action of digestive enzymes.

The carbohydrates found in the human diet are classified as monosaccharides or simple sugars, disaccharides or double sugars and polysaccharides which include many molecules of simple sugars. The monosaccharides are the simplest and smallest carbohydrate molecule requiring no digestion. They are absorbed from the intestine directly into the bloodstream.

Many of the simple sugars differ in the arrangements of the groupings about the carbon atoms, but the ones that should be noted are glucose, fructose and galactose. These three simple sugars are found in foods and are formed from other carbohydrates in the process of digestion.

Glucose, also known as dextrose, grape sugar or corn sugar, is abundant in fruits and vegetables. It is the form of carbohydrate to which eventually all other forms are converted for transport in the blood and utilization by the other tissues of the body.

Fructose, also called fruit sugar, is much sweeter than cane sugar but does not crystallize readily. It is found in ripe fruits and many vegetables and is responsible for the peculiar sweetness of honey.

Galactose is not found free in nature. It is derived from the double sugar, lactose, in the process of digestion.

The disaccharides or double sugars found in foods are lactose, maltose and sucrose. These double sugars are split to simple sugars by the action of specific enzymes in the digestive tract.

Lactose or milk sugar is the only common sugar not found in plants. It is produced only in the milk glands of nursing mothers, either animal or human.

Maltose, or malt sugar, is an intermediate product in the digestion of starch. It is also produced in the malting and fermentation of grains, thereby being found in beer and malted breakfast cereals.

Sucrose can usually be found by looking no further than the kitchen cupboard. Sucrose is the table sugar with which we are so familiar. It may be ordinary granulated, powdered or brown sugar. Many fruits, some vegetables and molasses also contain sucrose.

One of the sweetest forms of sugar, sucrose, is often consumed in greater quantities than any other sugar. Frequently, this is to the detriment of the teeth and health in general. High concentrations of sucrose have an irritating effect on tissues lining the digestive tract. It is also true that large amounts of sucrose tend to satisfy the appetite, thus reducing the intake of other more necessary foods.

The polysaccharides are formed by molecules that may be several hundred times as large as those of the sugars. These carbohydrates are not sweet, are insoluble in water and differ markedly as to digestibility and resistance to spoilage. Starch, dextrin and glycogen are forms of polysaccharides suitable for human digestion.

The chief value of starches lies in their ability to furnish energy for the body. In the plant, starch is the storage form of carbohydrate. Some common starches include potatoes, wheat, rice, arrowroot and corn. Before these starches can be used readily by the body they must be prepared properly. This is true because the starch of a grain lies encased in a protective covering of cellulose.

Cellulose, the most abundant organic compound in the world, is resistant to digestion. The skins of fruits, the coverings of seeds and the structural parts of edible plants are cellulose in nature. The indigestibility of cellulose is actually a major asset. Undigested cellulose fiber furnishes bulk that is necessary for the efficient expulsory action of the intestines.

By grinding or the application of heat and moisture in cooking the outer cellulose of a starch is ruptured, thus converting the starch to a form that can be more easily digested. As a starch is chewed in the mouth, the digestive process already begins to take place. Ptyalin, an active enzyme in saliva, transforms the original food into a more soluble starch, which in turn becomes the intermediate product known as dextrin. Finally the dextrin is changed into an entirely soluble product, maltose.

The length of cooking time affects how readily a starch may be digested by the body. It has been shown that cooking, if continued long enough, will carry the breakdown of starch to the dextrin stage.

The polysaccharide glycogen has been referred to as the animal starch. Liver and deep sea scallops contain the most of any familiar foods. Glycogen is the storage form of carbohydrate in the body.

Every type of carbohydrate eaten in the daily diet, whether simple sugars (monosaccharides), double sugars (disaccharides), or starches (polysaccharides), is used by the human body. Some carbohydrates meet the energy needs of the body and others are used to help build up more complex compounds used in metabolic reactions. Only very small amounts of carbohydrates are stored by the body. The amount stored in the liver, muscles and blood is so small that it would be hardly sufficient for the daily fuel needs of most individuals.

Good health cannot be maintained without proper nutrition. For this reason a person cannot ignore the need for a constant daily supply of carbohydrates in the diet. Just as proteins perform specific functions in the body, so also do carbohydrates.

If carbohydrates are not available in the diet in sufficient amounts, the body will convert protein to glucose in order to supply energy. This happens because the body's need for energy takes precedence over all else. Should this improper use of protein for energy occur, the body then suffers from a lack of essential tissue-building proteins, which can only be prevented by increasing the protein content of the diet.

When carbohydrates are supplied in the daily diet, the optimum utilization of amino acids for protein formation is to take place. When proteins and carbohydrates are eaten in the same meal, not only is protein best utilized by the body but the nitrogen balance is improved.

The presence of carbohydrates is also necessary for normal fat metabolism. Without sufficient carbohydrates, larger amounts of fats are used for energy faster than the body can take care of the acidic intermediate products, and loss of weight will take place.

The carbohydrate sugars react chemically within the body to become vitally important constituents of numerous compounds. These compounds are present in the heart valve, skin, bone, cartilage and tendons. In the liver glucuromic acid, formed in part of glucose, combines with chemicals and toxins, converting them into a form that may be excreted.

Other examples of vital functions of these compounds include: the formation of the matrix of connective tissue, the prevention of blood clotting, the resistance of the body against infection, and the transfer of genetic characteristics of the cell.

Lactose and glucose cannot be forgotten in any explanation of the functions of carbohydrates. Lactose, a double sugar found in milk, enhances the absorption of calcium, besides promoting the growth of desirable bacteria in the small intestine.

Glucose is the major source of energy for not only the body, but the brain and nervous tissue. It must be supplied regularly for the functioning of these tissues. If glucose is not supplied to or manufactured by the body, the brain will then call upon by-products of the breakdown of stored fat to provide its fuel.

Even the less digestible carbohydrates are utilized by the body. Cellulose, as mentioned before, gives invaluable aid to normal elimination by adding bulk to the diet and stimulating the gastrointestinal tract.

Whatever the type of carbohydrate that is eaten in the daily diet, it must still be reduced to one form, glucose, before it can be used by the body. That is the purpose of carbohydrate digestion. This digestive process takes place first in the mouth when a food is chewed. It continues in the stomach after food is swallowed, and from there proceeds to the jejunum, the middle part of the small intestine, which is the principal site of carbohydrate digestion.

The less digestible carbohydrates continue through the intestinal tract and are excreted in bowel movements. The remaining carbohydrates are absorbed in the jejunum to be used elsewhere in the body. Because of this process the actual amount of carbohydrate available to the body is the difference between the total amount contained in a food and the amount of fiber found in the food.

Glucose is the most important and versatile carbohydrate used by the body. Before a carbohydrate can be transported in the blood to the tissues it must eventually be reduced to glucose by the body. When we talk about what happens to a carbohydrate after digestion we are essentially talking about how the body will use the glucose it has received.

Glucose metabolism, by which substances are broken down to yield energy, goes hand in hand with the metabolism of fats and proteins. Just as the body can convert proteins and fats to glucose if necessary, glucose itself can be converted to fatty acids, glycerol and certain amino acids.

This fact is an excellent example of the interdependence of the various nutrients. The lack of any one of them will affect the total metabolism.

The story of carbohydrates is not completed with the digestion of such foods or its absorption into the body. What remains is to know at what point and how a carbohydrate will be used as energy by the body.

After absorption from the small intestine the monosaccharides are carried by the blood to the liver. The liver releases glucose at the exact rate needed by the body. This mechanism is controlled largely by insulin from the pancreas. Also influencing reactions in the liver are hormones secreted from the adrenal, pituitary and thyroid glands.

Glucose is carried to where it is needed in the body by the blood. The blood glucose level is maintained within narrow limits, rising shortly after a meal but dropping within a few hours. Circulating blood glucose is absorbed by the muscles and other tissue cells. The muscle glycogen is the reserve fuel for muscular energy.

Energy is not produced from glucose in a single reaction. The breakdown of glucose is a complex chemical reaction. Numerous steps occur, precipitated by specific enzyme reactions, until a complex glycogen-phosphorus compound is produced that can break down with almost explosive speed when the nerve impulses of the body demand muscular action.

Actions that we constantly take for granted, such as turning the pages of this book, require muscular activity. It is imperative that our bodies have carbohydrates to supply them with fuel for such actions. Each gram of carbohydrate yields approximately 4 calories. The fuel factor of calories produced by carbohydrates is based on the level of absorption of these foods.

However, the body can store and utilize only so much carbohydrate. Surplus carbohydrates are transformed into fat and stored as fatty tissue. When more carbohydrates are eaten than the body needs for daily use, obesity can result.

Distinctions must be made between valuable carbohydrate foods and carbohydrate foods that may be harmful to the health if they take precedence in the daily diet. Most carbohydrate foods contain more than one nutrient.

They are valuable foods to include in the diet since the body can use them in numerous ways, as determined by the specific needs of the body at any given time. Wheat, corn, rice, enriched breads and whole grain cereals contain varying amounts of proteins, B complex vitamins and iron. Fruits and vegetables are valuable not only for the digestible carbohydrate and cellulose they contain but also because of their vitamin and mineral content.

Sweets such as candy, honey, jellies, molasses and soft drinks contain little, if any, other nutrient. They are referred to as "empty calories" because they contribute nothing except calories to the diet. This type of carbohydrate should be permitted in the diet only in restricted amounts, or better still, not at all.

That each nutrient fulfills a vital role in maintaining life is becoming more and more apparent. That the quality of nutrients must take precedence over quantity is something that must be kept uppermost in everyone's mind. The chart of recommended dietary allowances is a valuable reference guide to consult before planning for daily meals. No other practical guide shows quite as clearly how to integrate carbohydrate foods in the diet—not as separate entities but as part of the combined daily intake along with proteins and fats.

FATS

Fats are as important a form of stored energy for animals as carbohydrates are for plants. During the past century there has been an increase in the per capita use of fats in the United States. It is almost impossible to estimate how much is really consumed since the daily intake of fat varies so widely among individuals and regions.

Fats actually make the daily mealtime more enjoyable. Imagine a meal without butter or margarine on bread, dressings on salads, sauces on meats or vegetables. Psychologically, fats have a great value in the daily diet.

Realistically, fats serve multiple purposes in the diet. Besides making meals more palatable and adding to the flavor of foods, fats also retard the rapid development of hunger. They do this because fats are digested slowly and stay in the stomach longer than other nutrients. Some fats are carriers of the essential fatty acids and others aid in the transport and absorption of fat-soluble vitamins.

Adipose tissue, which consists primarily of stored fat, surrounds vital body organs and is interlaced throughout muscle tissue. This fatty tissue helps to hold the body organs and nerves in place and protects them against physical injury. Other fat layers in the body act as insulators, reducing the loss of body heat and maintaining body temperature. Fats and oils also have some value as a lubricant for the gastrointestinal tract.

Before going further it should be said that while fats are valuable to the body, too great a fatty deposit can be dangerous. Obesity may result, placing an unnecessary burden upon the heart and other organs. The high consumption of fat in the United States may predispose to obesity, not necessarily because fat foods produce fat, but because people who eat a high fat diet can so easily overstep their caloric needs.

Each gram of fat supplies 9 calories. This is more than twice the amount of energy supplied by each gram of carbohydrate. Fats rather than carbohydrates supply up to two-thirds of the total energy of the cells.

The major food sources that supply fat to the diet are butter, margarine, lard, vegetable oil, salad dressing, the visible fat of meat, the skin of chicken and the invisible fat found in cream, homogenized milk, milk products, egg yolk, meat, fish, nuts, olives, avocados and whole wheat cereals. The so-called invisible fats represent about three-fifths of the fat in the American diet.

Like carbohydrates, fats are organic compounds of carbon, hydrogen and oxygen, but there the resemblance ends. The fatty acids and glycerol, composed of carbon, hydrogen and oxygen, are insoluble in water and greasy to the touch.

Fats that are referred to as lipids include oils, which are fluid at room temperature and fats, which are usually solid. All of these lipids or fats, as they are more commonly called, have similar chemical structures. These lipids are generally classified in three groups.

The first group, simple lipids, includes the triglycerides which account for about 98 percent of the fats in food and over 90 percent of the total fat in the body. Triglycerides are formed when the fatty acids react with the carbon alcohol, glycerol.

Notable in the second class of lipids, the compound lipids, are the phospholipids. Any lipid containing phosphorus is included in this group. These lipids function in maintaining the structural integrity of the cells rather than as fat stores. Phospholipids are contained in all cells of the body. The brain, nervous tissue and liver are especially rich in them.

Phospholipids are essential to the digestion and absorption of fats. Large concentrations of compound lipids combine with protein in cell membranes where they act as a liaison between fat-soluble and water-soluble substances that facilitate the passage of fat in and out of the cell.

The derived lipids comprise the last group of the lipid classification. Cholesterol and fatty acids are part of this group. Cholesterol is present in almost all body tissues. Like the phospholipids, it combines with protein to enable fats to be transported to the cells. Cholesterol is necessary for the buildup of provitamin D, bile salts, adrenocortical and steroid sex hormones.

Only animal foods furnish cholesterol. The body, however, continually manufactures cholesterol regardless of dietary intake. A person who eats no eggs or organ meats probably ingests not more than 200 mg. of cholesterol daily. Each egg yolk adds about 275 mg. of cholesterol.

The types of fatty acids in fats are responsible for differences in texture, melting point and other characteristics. Some fatty acids are saturated, meaning that each carbon atom in the long chain molecule carries all the hydrogen atoms possible. Fats formed from such fatty acids are usually solid at room temperature. As a rule, the fat of plant-eating animals is harder than that of carnivorous animals, and land animals have harder fats than aquatic animals.

An unsaturated fatty acid is one in which one or more carbon atoms fail to carry all of the hydrogen atoms possible. The unsaturated fatty acids are found mostly in oils from seeds and fish or in softer fats. Some of the more highly unsaturated fatty acids cannot be synthesized or built up within the body from other fatty acids. These are known as essential fatty acids because they are necessary for growth and health. The essential fatty acids must be supplied in food.

Linoleic acid and arachidonic acid are both essential fatty acids and polyunsaturated fats. Linoleic acids are found in all vegetable fats, except coconut oil. Corn, cottonseed and soy oil are good sources. Essential fatty acids are constituents of phospholipids that form cellular membranes and appear to play a role in regulating cell permeability and the transportation of lipids in the circulation.

When fats are referred to as being polyunsaturated it means simply that these fats have a high proportion of fatty acids with two or more double bonds. Vegetable fats are polyunsaturated fats which are used to make margarine and shortening. This is done by the process of hydrogenation.

Hydrogenation causes liquid fats to be changed to solid fats by the introduction of hydrogen to the double bonds of the carbon chain. Complete hydrogenation could produce a very hard fat. When the process is controlled, any desired consistency can be prepared. A serious disadvantage of hydrogenation is that it lowers the polyunsaturated fatty acid content of the fat.

Another reaction that is a typical characteristic of fats is emulsification; that is, fats are capable of forming emulsions with liquids. This happens during the homogenization of milk and the preparation of mayonnaise. In the body emulsification is essential for the digestion and absorption of fats.

Since fats are so susceptible to chemical changes they tend to spoil rapidly. Rancidity results when fats and oils are exposed to warm, moist air over a period of time. Oxygen can attack the double bonds of polyunsaturated fatty acids, forming toxic substances. Changes in odor and flavor become quickly apparent. Precautions should be taken to store fat-containing foods at low temperatures to guard against rancidity.

Foods fried in fat may undergo still another undesirable chemical reaction if the fats are heated excessively. Fats, when fried at too high a temperature, decompose to produce a compound that may be irritating to the intestinal mucosa. On the other hand, when the frying temperature is too low, foods absorb excess amounts of fat, thus lengthening the time required for digestion.

Fats naturally tend to be digested slower than carbohydrates and proteins. Unlike carbohydrates or proteins, fats require special digestive action before absorption. The end products of the chemical changes that occur in digestion must be carried in a water medium, blood and lymph, in which fats are not initially soluble.

When fats leave the stomach and enter the duodenum of the intestine their presence stimulates the intestinal wall to secrete cholecystokinin. This hormone in turn stimulates the contraction of the gall bladder, forcing bile into the common duct and ultimately into the small intestine. Bile emulsifies all fats in the intestines. In doing this, fats are readied for optimum enzyme action. After fats are split by enzymes they can be presented as fatty acids, glycerol, mono-, di-, and triglycerides for absorption.

Most of the absorption of fats occurs through the intestinal mucosa of the jejunum. Normally about 95 percent of dietary fats and 80 percent of dietary cholesterol are absorbed. During absorption, that is, when fats have penetrated the intestinal mucosa, glycerol and fatty acids re-combine with a small amount of protein to form microscopic particles of fats called chylomicrons. Only then can they enter the lymph, part of the circulatory system, and be carried to the liver. The liver is the key organ in the regulation of fat metabolism.

Once again it seems that the discussion has come full cycle to the process of metabolism. This culmination of events is only natural in any discussion of the essential need for all nutrients. Metabolism, or simply how the body uses foods in the daily diet can only reinforce the fact that one nutrient should not take precedence over another. All are equally essential for good health and nutrition.

The body uses fats in much the same way it uses carbohydrates. That is to say, both fats and carbohydrates are mainly energy foods.

The breakdown of fats resembles the breakdown of glucose. Both processes consist of a long series of many of the same chemical reactions.

Fats are first converted to fatty acids and glycerol, primarily in the liver. Glycerol is converted to glucose or glycogen which can then be broken down to yield energy by the usual process of carbohydrate catabolism. Fatty acids are converted to ketone bodies. Ketone bodies are used to produce energy in muscles and all other cells.

All fats are not used by the body for energy. Fats are also used by the cells to build necessary compounds needed by the body. Some fats are stored in the liver and in adipose tissue. Adipose tissue, which consists mainly of triglycerides, has the necessary enzymes to be able to constantly produce and release new fat to meet the body's needs.

The Food and Nutrition Board has not set precise recommendations for either the quantity or type of fat that should be included in the normal diet. However, it should be remembered that many fats such as oil, lard, butter, margarine, bacon and salad dressings, are very concentrated sources of calories. A small amount of them goes a long way toward meeting the body's needs. If these fats are not used in moderation a person can very rapidly take in more fat than his body needs and become overweight.

Ordinarily, the body can obtain the fat it needs for energy from foods which also supply the body with necessary proteins. Protein foods take the place of first importance in the diet. Interestingly enough, protein alone is never the sole nutrient found in a food. Protein-rich foods contain either carbohydrate or fat. Nature has wisely provided foods with a combination of nutrients, since the life-sustaining protein cannot be used properly without other nutrients. It stands to reason that our diets should never be deficient in the three essential nutrients: protein, carbohydrate, and fat. In order to place emphasis on specific ways in which good quality nutrients can be used at mealtime, the end of the chapter lists a wide variety of menu plans.

THE BALANCE BETWEEN ACID AND ALKALI

Protein also helps maintain the body's balance between the acid and alkaline state. Parotid glands assist the digestion of protein when the acid-alkaline balance is upset.

The digestion and assimiliation of all classes of protein are almost stopped if the parotid glands are dehydrated.

Alkalinity and Acid

Alkalinity is necessary to the life process. Seeds will not germinate unless first softened and alkalinized by water. Conception cannot occur if a mother's cells are not alkalinized. This process falls under the Law of Polarity—the earth and all contained therein is a manifestation of the fusion of positive and negative forces, with a line of balance in the center. In the body there must be a balance between alkali and acid.

The pituitary gland seems to be the balancer in the event of alkaline-acid imbalance, while the other glands have subsidiary duties.

Other activities of the pituitary gland include: regulating fluid intake and output, stimulating lactation and the thyroid gland, and sexual development. The pituitary gland is also involved in the stimulation of growth and of metabolism (digestion and assimilation). It starts the menstrual cycle and helps steady gaseous metabolism.

An imbalance in the acid-alkaline content occurs when one eats food of such poor quality that the subsidiary glands become overtaxed and congested. When this happens the entire body mechanism is thrown off track. Disappointment, shock, or emotional upset should never unbalance a properly nourished person.

One of the chief functions of the thyroid gland is to prevent the body from absorbing the toxic wastes from proteins. To achieve this, the thyroid employs iodine. With its alkaline basis, iodine has the power to neutralize acid. Animal proteins contain a high quantity of toxic matter, and the ingestion of too much of these proteins might well cause a thyroid imbalance. That is not to say that excluding animal proteins from the diet is the simple solution. Too much of any one food can be harmful. The knowledge of both the quantity of protein a person must have to replace tissue, supply heat, and maintain blood sugar level and the capacity to handle protein is important.

A person's individual make-up determines the amount of a given food he can handle. Proper digestion and assimilation can make the difference between health and illness.

The largest gland in the body is the *liver*. While it stimulates peristalsis, the major function of the liver is to secrete the bile which assists digestion in the intestines.

The wavelike contractions along the alimentary tract, which force its contents onward, are described by the term *peristalsis*. Another necessary function of the liver, with the help of the spleen, is to break down dead blood cells and continuously detoxify the blood.

The *pancreas gland* releases a digestive fluid containing ferments, which break down all classes of food. The chief regulate fat and sugar metabolism. The pancreas is also a function of this six-inch organ is the secretion of insulin to detoxifier.

The *adrenal glands* provide immunity against infection while attending to the healing of cuts and wounds.

The *kidneys*, like the alimentary tract, the skin, and the lungs, are chiefly concerned with excretion. All the excretory processes are stimulated by the adrenals. Their energizing qualities affect mental, emotional and physical well-being.

As the body's defender, the kidneys can separate iron tablets, animal hormones, drugs, undigested proteins, and other harmful substance in the fluid of the body and eliminate them through the urine. The urine of a healthy person is 95 percent water. Also present in urine are inorganic salts and organic waste material including urea, creatinine and uric acids. Over-accumulation of these harmful substances can result in permanent body damage.

Toxemia, a dispersement of toxins (poisonous substances) throughout the body, will deplete the adrenals quite rapidly. This condition allows easy detection of any imbalance of these glands. Muscular weakness, hot flashes and sudden chills from a circulatory origin, dryness of the mouth, nervous tension, a sense of overall weakness, and a fall in blood sugar are all symptoms of glandular imbalance.

The *parotid glands* and the *salivary glands* are two lesser known glands located in front of each ear. The parotids are the largest of the mucous glands. The salivary glands produce the fluids necessary for digestion. Saliva, the colorless fluid produced by the combined secretions of these glands, contains digestive ferments and enzymes.

One of these ferments, ptyalin, changes starch into dextrin and sugar.

Rapid loss of the body's acid-alkaline balance occurs when the paratids fail to secrete at the same time the digestive glands in the stomach are unable to produce hydrochloric acid in sufficient quantities to change protein into amino acids. This condition is called alkalosis or increased alkalinity of the body fluids. Acidosis is excessive acidity and, since most people are far too quick to judge this to be the cause of their trouble, alkalosis is the more difficult to overcome. Antacids worsen alkalosis.

It is difficult to know whether the parotids or the pituitary has the most influence over the sex organs, because this is an area in which the two function parallel with each other.

Let's now examine the parotid gland and its relationship to the common symptoms of acid-alkaline imbalance. Since a negative and a positive force form every chemical compound, parotid secretions, which are alkaline and therefore negative, would not be able to function without gastric acids, which are positive.

The body can get only the percentage of nourishment which the parotids are secreting of their normal quota. Distress would not be produced under a low percentage of functional capabilities, but there would be deterioration of muscle and other tissues. However, decreased peristaltic action and near paralysis of the bowels results from deficiencies of liver fluids. Stagnation and decay of undigested food in the intestines is a greater problem of parotid disturbance than slow starvation.

Just as nature produces moss to sweeten a stagnant pool of water, the body creates a fungus to absorb the decaying waste matter which accumulates in the organs. This fungus is a protective organism which attaches itself to part of the body—the ears, eyes, liver, tonsils, heart, mastoids—where accumulations of toxic waste threaten to destroy the cells. Having done its work, the fungus disappears.

The *parathyroid glands,* located behind the thyroid gland, have as their major function the task of keeping calcium in the bloodstream. Normally four in number, these small glands also destroy waste products, especially protein by-products which affect the nervous system.

Whenever the parotids and adrenals are out of commission, the parathyroids take over the depleted glands' functions until they become active and back in balance.

Acid: Keynote of Energy

Working together, acids and alkalies are an integral part of the life process. Alkalies hold back the driving force of the positive acids. Without acids, cells would stop recreating new cells and your body would quickly decline. Conversely, without alkalies, the stomach would digest itself, and the unchecked acids would burn up the body.

The pituitary gland directs activities during the constant state of flux which accompanies the re-creation of cells. This busy organ is thrown off balance after any shock hits the system, be the experience heavy exercise, quick temperature change, mental stress, or even a meal. Time out for reestablishing that balance is a must for this overworked organ. Several hours generally pass before a state of near balance is reached.

Heartburn, certain gastric conditions, and arthritis, among others, are the symptoms of acid-alkaline imbalance, but do not necessarily mean that you are suffering from acidosis. They could possibly be warning you of alkalosis. Never prescribe for or diagnose yourself. Your family doctor can spare you pain and lingering illnesses, so don't make matters worse for him.

MENU PLANNING

Food, the basis of life, is composed of many nutrients. Proteins, carbohydrates, fats, minerals, water and vitamins are nutrients that are found in foods that nourish the body.

Each type of food offers certain benefits. A good balanced diet can be found only in using a variety of foods. Therefore, it is necessary to know the nutritive content of foods, what are the best sources of the various nutrients and how to combine them into a healthy, balanced diet. No one food does everything, and all foods have something to offer.

The more varied your diet the better off you will be at

all ages of life. Regardless of age, everyone needs the same nutrients, but in different amounts. The man doing physical labor needs more energy than the man who sits at a desk all day.

In 1940 the Food and Nutrition Board of the National Research Council was organized in the United States to guide the government in its nutrition program. The Board made a careful review of research on human requirements for the various nutrients. This led to the publication of the Recommended Dietary Allowances in 1943 known as the R.D.A. From time to time new research has published revisions of these standards.

"A Daily Food Guide" as set up by the R.D.A. provides a foundation for a day's meals and includes food choices which permit flexibility for seasonal, regional and budgetary considerations.

The minimum number of servings for the adult of each of the four food groups will be used as a basis for planning the menus that follow. The size of each portion will obviously need to be modified for preschool children, active teenagers, pregnant women and senior citizens to provide the correct amounts of the various nutrients.

Variety in meal planning is the sum total of many kinds and classes of foods served in pleasing combinations with a careful mixture of soft and crisp foods, bland and sharp flavors, and hot and cold dishes. It insures better nutrition and enhances the interest in the meal.

A common error in meal planning is emphasizing one type of food to the exclusion of others. Therefore, variety means selecting foods each day from each of the basic four food groups.

The first appeal to the appetite is through the eye. Food must look "good enough to eat," and attractive color combinations are important. Sometimes just a sprig of parsley will dress up a plate.

Texture variation is equally important. Foods that are all soft lend no interest. Yet, on the other hand, children do not like foods that require long amounts of chewing. Also, older people may not be able to chew foods that are tough. Taste appeal depends on blending of sharp and bland flavors.

The climate and season of the year requires some consideration. On a hot summer day a cold soup is much more appetizing, while on a cold wintry day a hearty soup is most welcome. Meals can also be varied by having a heavy breakfast one day and a lighter another day. This can also be true with lunch and dinner.

In planning the menus a high quality and variety of foods are used to insure the minimal daily requirements plus additional nutrients. Therefore, all menus are planned based on the Basic Four Food groups as outlined by the Recommended Daily Allowances.

Meat has been used in some dishes because, as we mentioned earlier, we do not believe in drastically changing a person's eating concepts. Rather, we believe in a gradual program of selecting alternative foods.

The menus will be broken down with distinctive notations being given to the various age levels as well as needs. Also included are various suggestions for changes and substitutions.

MENUS-1

Breakfast

Orange juice
Oatmeal with honey and milk
Sausage and whole wheat toast
Milk for children
Minted tea or decaffeinated coffee for adults

Lunch

Chopped egg salad sandwich on rye bread
Carrot strips
Stewed pears
Milk for everyone

Dinner

Meat loaf made with kasha as a filler
Oven browned potatoes
Spinach timbales
Celery curls
Whole wheat bread and butter
Blueberries or peaches
Milk, minted tea or decaffeinated coffee

Snack

Yogurt or apple or milk

Using the above menu for the family it is necessary that the young child has milk at each meal while the teenager not only has milk at all meals but also has milk as a snack. The older person may have milk for lunch and dinner, or as a snack. The pregnant woman should have milk for lunch and dinner and also as a snack. For a very young child and also for senior citizens you may want to eliminate the sausage and possibly the toast for breakfast. Use fresh fruit whenever it is available and within the price range of the family such as blueberries or fresh peaches in the summer.

2

Breakfast

Half a grapefruit
French toast and syrup or honey
Milk
Herbal tea or decaffeinated coffee

Lunch

Cream of tomato soup
Toasted cheese sandwich on rye bread
Carrot strips
Tangarine sections
Milk, herbal tea or decaffeinated coffee

Dinner

Crunchy corned beef
Potato salad
Marinated green beans
Salad greens
Whole wheat breadsticks and butter
Cantaloupe
Milk, herbal tea or decaffeinated coffee

Snack

Milk, dried apricots or buttermilk

The young child will have milk three times a day, while the teenager needs four glasses of milk daily. The adult may choose milk for lunch and snack or for dinner. The pregnant woman needs at least three glasses of milk daily. The young child may desire celery curls or pieces of raw turnips instead of salad greens.

3

Breakfast

Tomato juice
Fried cornmeal mush with honey
Strips of well-cooked bacon

Lunch

Homemade vegetable soup and rye sticks
Deviled eggs
Fresh plums
Milk, herbal tea or decaffeinated coffee

Dinner

Fish sticks
Corn on the cob
Acorn squash seasoned with maple syrup
Sliced tomatoes and onions
Cracked wheat bread and butter
Vanilla pudding with strawberries
Milk, herbal tea or decaffeinated coffee

Snack

Yogurt or fresh pears or milk

Children may not relish fried mush as well as cooked hot mush, served as a cereal, which is nourishing and tasty with honey and milk. Instead of deviled eggs at lunch for small children, they may like having a hard-cooked egg in the shell that they can break open and shell themselves. This gives youngsters a feeling of "belonging" and having a part in the daily food preparation. Instead of corn on the cob, which children enjoy, the senior citizens who have trouble with their teeth would enjoy having the corn cut off the cob. This can be done before the corn is cooked or before it is served.

Breakfast

Stewed prunes with a slice of lemon
Whole grain cereal with honey and milk
Poached egg on toast
Milk, herbal tea or decaffeinated coffee

Lunch

Creamed cheese–nut sandwich on raisin whole wheat bread
Orange or grapefruit sections
Milk, herbal tea or decaffeinated coffee

Dinner

Pan-broiled liver with bits of bacon
Succotash made from fresh corn and beans
Glazed carrots
Mixed green salad
Corn bread and butter
Honeydew with lime wedges
Milk, herbal tea or decaffeinated coffee

Snack

Milk, caramel custard or orange

Organ meats are one of the richest sources of protein, and we suggest you use them in some form at least once a week. They are usually less expensive. If you are trying to get more milk in your diet the caramel custard for a snack is a good way. This is enjoyed by all.

5

Breakfast

Strawberries with wheat germ and cream and honey
Sausage with waffles and honey
Milk—herbal tea or decaffeinated coffee

Lunch

Cheese fondue
Watercress salad
Pineapple wedges
Milk, herbal tea or decaffeinated coffee

Dinner

Spaghetti with meatballs, parmesan cheese
Broccoli spears
Tossed green salad
Garlic bread and butter
Assorted fresh fruit
Milk, herbal tea or decaffeinated coffee

Snacks

Buttermilk or Jell-O or milk

Youngsters usually like waffles and with honey on them. They make a special breakfast for young and old. Strawberries with wheat germ and honey can be a real treat. Cheese fondue can be so good, but children may prefer a cheese sandwich to cheese fondue. Spaghetti is one of the favorite dishes of children, and along with green vegetables and salads meets the daily nutrient needs. Some raw, chopped-up spinach added to salad not only adds much color but also enhances the vitamin A content of the daily requirement. Jell-O may be a favorite snack for all. But be sure the teenagers have the extra cup of milk.

6

Breakfast

Orange or cranberry juice
Scrapple with molasses
Oatmeal biscuits
Milk, herbal tea or decaffeinated coffee

Lunch

Peanut butter sandwich on whole wheat banana bread
Waldorf salad
Rice pudding
Milk, herbal tea or decaffeinated coffee

Dinner

Tuna casserole with whole wheat noodles and peas
Spinach soufflé
Carrot-raisin salad
Rye bread and butter
Prune whip
Milk, herbal tea or decaffeinated coffee

Snack

Milk, herbal tea or decaffeinated coffee

Scrapple—a real old-fashioned Pennsylvania dish—is made with cornmeal and served with molasses. It is tasty and quite nourishing. Banana bread has such a delicious flavor, especially when it is served with peanut butter. Waldorf salad, which consists of apples, celery, nuts or raisins helps to fulfill the protein need. Your family may not all like tuna casserole and prefer it as tuna salad or just a serving as chunks of tuna and peas and noodles served separately. Milk with each meal is a necessity for teenagers and pregnant women. The young child and adult can get along with two glasses of milk. Casserole dishes add much protein to the diet through the use of milk in preparing them. They are a means of stretching the food dollar through the use of cheese and leftover bits of meat.

Breakfast

Grapefruit juice
Spanish omelet
Whole wheat coffee cake and butter
Milk, herbal tea or decaffeinated coffee

Lunch

Baked noodles, cheese and tomatoes
Hearts of lettuce—Italian dressing
Watermelon
Milk, herbal tea or decaffeinated coffee

Dinner

Barbecued spareribs
Oven-browned potatoes
Spinach timbales
Coleslaw
Cornbread and butter
Lemon snow
Milk, herbal tea or decaffeinated coffee

Snack

Yogurt or pears or milk

Children may not like spinach omelet, but you can give them scrambled eggs, thereby not having to change the market order and keeping everybody happy and well nourished. Barbecued spareribs are a real treat for children because they can take them in their hands. What we call "finger foods" are always a pleasure for children. Yogurt is an excellent snack for adults while children can have milk or pears. Of course it is necessary that teenagers and nursing women have four glasses of milk a day. The pregnant woman needs at least three glasses of milk daily. One glass can always be taken as a snack.

Breakfast

Orange sections
Buckwheat pancakes and maple syrup
Strips of bacon
Milk, herbal tea or decaffeinated coffee

Lunch

Black bean soup
Chef's salad
Cup custard
Milk, herbal tea or decaffeinated coffee

Dinner

London broil
Parsleyed potatoes
Broiled tomatoes
Cabbage salad
Bran muffins and butter
Milk, herbal tea or decaffeinated coffee

Snack

Strawberry yogurt or purple plums or milk

Chef's salad usually has strips of cheese, ham and turkey. Any one of them can be omitted if you need to stretch the budget. Flank steak is used for London broil and is just as nutritious as an expensive cut of beef and very delicious. Cabbage is very high in vitamins and less expensive than some of the other salad greens. Parsley will dress up any potato and is nutritious as well as tasty.

9

Breakfast

Honeydew melon
Bran flakes with wheat germ and milk
Milk, herbal tea or decaffeinated coffee

Lunch

Soybean square
Eggplant and tomato
Whole wheat wafers and butter
Baked apple
Milk, herbal tea or decaffeinated coffee

Dinner

Steamed frankfurters
Baked beans
Raisin-apple-carrot salad
Oatmeal muffin and butter
Orange slices
Milk, herbal tea or decaffeinated coffee

Snack

Tomato juice or dates or milk

All youngsters as well as grownups like frankfurters and baked beans sometimes. Apple, raisin and carrot salad is a good combination and makes a varied dish with color and vitamins and iron.

10

Breakfast

Pineapple juice
Eggs Benedict
Bran muffins and butter
Milk, herbal tea or decaffeinated coffee

Lunch

Peppers stuffed with hamburger
Green bean salad
Rye bread and butter
Peanut butter cookie
Milk, herbal tea or decaffeinated coffee

Dinner

Sautéed cod fish cakes
Broiled acorn squash
Stewed tomatoes
Pear–cottage cheese salad
Three flour bread and butter
Apple-sauce
Milk, herbal tea or decaffeinated coffee

Snack

Milk, Triscuit or buttermilk

Cod fish cakes are not only tasty, but also among the most economical fish dishes today. Acorn squash broiled with honey inside of it and nut meats is a very tasty autumn dish and very high in vitamin A.

11

Breakfast

Broiled grapefruit
Grilled sausage
Buttermilk pancakes and molasses
Milk, herbal tea or decaffeinated coffee

Lunch

Eggs baked in whole wheat toasted baskets
Sliced tomatoes and cucumbers
Cottage pudding
Milk, herbal tea or decaffeinated coffee

Dinner

Cheese soufflé
Swiss chard
Glazed carrots
Tomato aspic
Blueberry muffins and butter
Ambrosia (coconut and fruit cup)
Milk, herbal tea or decaffeinated coffee

Snack

Apricot nectar or milk or apple

Buttermilk pancakes and molasses are a very healthy beginning for a cool, brisk day. Molasses adds much iron to the diet. Many times when children and adults have difficulty in getting sufficient milk in their diet, a soft pudding or custard has a great appeal. Swiss chard is as rich in vitamin A as spinach, and in many markets is much cheaper, also it adds variety to the diet. Coconut turns ordinary fruit cup into Ambrosia.

CHAPTER THREE

UNDERSTANDING FOOD VALUES

This chapter is intended to aid the reader in the practical application of the information presented so far.

The emphasis is placed upon unprocessed and "minimally" processed foods, since they are generally higher in protein and other nutrients than processed food.

The foods are listed in descending value of usable protein within each category. For each food you are given both the total grams of protein for an average serving and the number of usable grams of protein the food provides.

Seafood

Many people on a vegetarian diet will include seafood in their diets. Fish is close to meat in protein content and often superior in protein quality. It is also particularly high in the amino acid lysine.

Fish has been a successful staple diet for many people, especially in the Orient. Since rice is deficient in lysine and isoleucine, fish makes an excellent supplement to it.

SEAFOOD

Average Serving, 3½ oz or about 100 grams	*Total Protein*	*Usable Protein*
1. Tuna, canned in oil, drained	24 grams	19 grams
2. Mackerel, Pacific	22	18
3. Halibut	21	17
4. Humpback salmon	20	16
5. Swordfish	19	15
6. Striped bass	19	15
7. Rockfish	19	15
8. Shad	19	15

Average Serving, 3½ oz or about 100 grams	*Total Protein*	*Usable Protein*
9. Shrimp	19	15
10. Sardines, Atlantic, in oil	21	14
11. Carp	18	14
12. Catfish	18	14
13. Cod	18	14
14. Pacific herring	18	14
15. Haddock	18	14
16. Crab	17	14
17. Northern Lobster	17	14
18. Squid	16	13
19. Scallops, 2 or 3	15	12
20. Flounder or Sole	15	12
21. Clams, 4 large, 9 small	14	11
22. Oysters, 2 to 4	11	9

Dairy Products

Milk is only 4 percent protein. Eggs are only 13 percent protein. But these figures are misleading. You will remember from the protein quality discussion earlier that both of these foods contain the highest quality protein. The Net Protein Utilization (NPU) of milk is over 80 percent, and that of eggs is a startling 94 percent. Your body can use almost all of the protein it takes in from these two sources. Two cups of milk will supply more than one-third of your daily protein allowance.

The NPU range of dairy products is from 70 to 94. Most of them fall between 70 and 83. Dairy products are also high in calcium, so if dairy products are eliminated from your diet you should find another source of calcium.

Like fish, dairy products are very high in lysine. Consuming them can help you make up for lysine's deficiency in plant protein sources. But some dairy products are not good sources of protein. Cream, sour cream, and cream cheese contain too many calories for the protein they have. Butter contains no protein at all.

NUTRITIVE VALUE OF EGGS

Dietetic characteristic	Amounts per 100 g (excluding shell)		
	Units	*Yolk*	*White*
Water	g	49	87.8
Ash	g	2.0	0.3
Crude protein	g	16.0	10.9
Gross energy	g	347	50.9
Carbohydrate	g	0.6	0.8
Fats: Total	g	30.6	0.2
sat. fatty acids	g	10.3	—
oleic acid	g	14.7	—
linoleic acid	g	2.5	—
Calcium	mg	141	9.0
Phosphorus	mg	569	15
Iron	mg	5.5	1.1
Sodium	mg	52	146
Potassium	mg	98	139
Thiamine	µg	224	4.4
Riboflavin	µg	440	271
Niacin	µg	44	110
Vitamin A	i.u.	3,400	0

Calculated from original data of Brooks and Taylor (1955) and Watt and Merrill (1963)

DAIRY PRODUCTS

Average Serving	*Total Protein*	*Usable Protein*
1. Cottage cheese, 6 tbsp.		
creamed	14 grams	11 grams
uncreamed	17	13
2. Egg white, dried, or powdered ½ oz.	11	9
3. Milk, nonfat dry solids, tbsp. 1 oz.	10	8
4. Parmesan cheese, 1 oz.	10	7
5. Milk, skim, whole or buttermilk, 1 cup	9	7

Average Serving	Total Protein	Usable Protein
6. Yogurt from skim milk, 1 cup	8 grams	7 grams
7. Swiss cheese, 1 oz.	8	6
8. Edam cheese, 1 oz.	8	6
9. Egg, 1 medium	6	5
10. Ricotta cheese, ¼ cup	7	5
11. Cheddar cheese, 1 oz.	7	5
12. Roquefort cheese or	6	4
blue mold, 1 oz.	5	4
13. Camembert cheese, 1 oz.	5	4
14. Ice cream, about ⅕ pint	5	4

Legumes (Dried Peas, Beans, and Lentils)

Many people overlook the important role of legumes in a vegetarian diet because they think of them as being dull, but some legumes actually have a protein content equal to or greater than that of meat.

Soups can be made from lentils, peas, black beans, and soybeans. It takes only a little imagination to turn these foods into a part of your daily diet.

All legumes are at least 20 percent protein. The highest, soybeans and mung beans, have NPU's of 61 and 57 percent, respectively.

LEGUMES

Average Serving, ¼–⅓ cup dry	Total Protein	Usable Protein
1. Soybeans or soy grits	17 grams	10 grams
2. Mung beans	12	7
3. Broad beans	13	6
4. Peas	12	6
5. Black beans	12	5
6. Cowpeas (black-eye)	12	5
7. Kidney beans	12	5
8. Chickpeas	11	5
9. Lima beans	10	5

Average Serving, ¼–⅓ cup dry	*Total Protein*	*Usable Protein*
10. Tofu (Soybean curd), wet, 3½ oz.	8 grams	5 grams
11. Lentils	13	4
12. Other common beans	11	4

Nuts and Seeds

Nuts and seeds are as rich in protein as the legumes and their Net Protein Utilization (NPU) factor is often higher. Nuts are seldom given a place of any importance in the American diet, yet they contain important minerals. The Brazil nut, for instance, is important in a vegetarian diet because of its sulfur-containing amino acids which are rare in plant protein. Nuts are also a good source of fat essential in maintaining good health.

Some nuts are not considered good sources of protein because of the high ratio of calories to protein. These include pecans, chestnuts, coconuts, filberts. hazelnuts, macadamia nuts, almonds, pine nuts, and English walnuts.

Nuts and seeds tend to be high in tryptophan and sulfur, but they are generally deficient in isoleucine and lysine.

The following is a compositional breakdown of some nuts.

CHEMICAL COMPOSITION OF NUTS AND DRIED FRUITS

NUTS	*Water*	*Protein*	*Carbohydrate*	*Fat*	*Minerals*
Acorns	4.10	8.10	48.00	37.40	2.40
Almonds	4.90	21.40	16.80	54.40	2.50
Beechnuts	9.90	21.70	19.20	42.50	3.86
Brazil nuts	4.70	17.40	5.70	65.00	3.30
Butternuts	4.50	27.90	3.40	61.20	3.00
Candlenuts	5.90	21.40	4.90	61.70	3.30
Chestnuts (dried)	5.90	10.70	74.20	7.00	2.20
Chufa (earth almonds)	2.20	3.50	60.70	31.60	2.00

NUTS	Water	Protein	Carbohydrate	Fat	Minerals
Coconut	14.10	5.70	27.90	50.60	1.70
Filberts	5.40	16.50	11.70	64.00	2.40
Hickory nuts	3.70	15.40	11.40	67.40	2.10
Paradise nuts	2.30	22.20	10.20	62.60	2.70
Pecans	3.40	12.10	8.50	70.70	1.60
Pignons	3.40	14.60	17.30	61.90	2.90
Pignolias	6.20	33.90	7.90	48.20	3.80
Pistachios	4.20	22.60	15.60	54.56	3.10
Black Walnuts	2.50	27.60	11.70	56.30	1.90
English Walnuts	2.50	18.40	13.00	64.40	1.70
Water Chestnuts	12.30	4.00	50.00	1.20	1.77
Peanuts	7.40	29.80	14.70	43.50	2.25
Peanut Butter	2.10	29.30	17.10	46.50	2.20
Almond Butter	2.20	21.70	11.60	61.50	3.00
DRIED FRUITS					
Apples	26.10	1.60	62.00	2.20	2.00
Apricots	29.40	4.70	62.50	1.00	2.40
Pears	16.50	2.80	66.00	5.40	2.40
Peaches	20.00	3.15	50.00	.45	2.15
Prunes	22.30	2.10	71.20	—	2.30
Raisins	14.60	2.60	73.60	3.30	3.40
Currants	17.20	2.40	74.20	1.70	4.50

NUTS AND SEEDS

Average Serving, 1 oz.	Total Protein	Usable Protein
1. Pignolia nuts 2½ tbsp.	9 grams	5 grams
2. Pumpkin and squash seeds	8	5
3. Sunflower seeds (3 tbsp.) or meal (4 tbsp.)	7	4
4. Peanuts	8	3
5. Peanut butter	8	3
6. Cashews	5	3
7. Sesame seed (3 tbsp.) or or meal (4 tbsp.)	5	3

Average Serving, 1 oz.	*Total Protein*	*Usable Protein*
8. Pistachio nuts	5	3
9. Black walnuts	6	3
10. Brazil nuts	4	2

The size of the servings here is conservative. One ounce of peanuts will supply only 7 to 8 percent of your daily protein needs. A 10-cent bag of peanuts (about 1½ ounces) gives 10 to 12 percent.

Grain Cereals and Their Products

Cereals are not considered a source of protein in the United States, but they provide almost half the protein in the world's diet. Their percentage of protein content is not high.

As with other sources of protein, grains and cereals must be evaluated from two different aspects—the quantity and quality of the protein they provide.

There are wide differences in the quantity of protein among various grains. Wheat, rye, and oats, for example, have 30 to 35 percent more protein by weight than rice, corn, barley, and millet. Not only does the protein content vary from grain to grain, but it varies within one type of grain. Wheat can range from 9 to 14 percent protein. The wheat with the highest protein content is hard red spring wheat. Durum wheat, often used in pasta, has the second highest protein content at 13 percent.

NPU values of cereal range from the low 50s to the low 60s. There are some exceptions, however. Whole rice has a NPU of 70 percent, the same as meat. Next is whole wheat germ with an NPU of 67 percent. Oatmeal and buckwheat have NPU values of 66 and 65 percent. All these values are higher than most other vegetable protein sources and comparable to beef.

Many of the grains and cereals are deficient in isoleucine and lysine. Foods rich in these essential amino acids should be added to your diet.

GRAINS, CEREALS, AND THEIR PRODUCTS

Average Servings	*Total Protein*	*Usable Protein*
1. Wheat, whole grain hard red spring, ⅓ cup	8 grams	5 grams
2. Rye, whole grain ⅓ cup	7	4
3. Egg noodles, cooked 1 cup	7	4
4. Bulgur (parboiled wheat), ⅓ cup, or cracked wheat cereal ⅓ c.	6	4
5. Barley, pot or Scotch, ⅓ c.	6	4
6. Millet, ⅓ c.	6	3
7. Spaghetti or macaroni cooked 1 c.	5	3
8. Oatmeal, ⅓ c.	4	3
9. Rice, ⅓ c.		
a. brown	5	3
b. parboiled (converted)	5	3
c. Milled, polished	4	2
10. Wheat germ, commercial 2 level tbsp.	3	2
11. Bread, commercial 1 slice, whole wheat or rye	2.4	1.2
12. Wheat bran, crude 2 rounded tbsp.	1.6	0.9

FLOUR

One Cup of Flour	*Total Protein*	*Usable Protein*
1. Soybean flour, defatted	65 grams	40 grams
2. Gluten flour	85	23
3. Peanut flour, defatted	48	21
4. Soybean flour, full fat	26	16

One Cup of Flour	Total Protein	Usable Protein
5. Whole wheat flour or cracked wheat cereal	16	10
6. Rye flour, dark	16	9
7. Buckwheat flour, dark	12	8
8. Oatmeal	11	7
9. Barley flour	11	7
10. Cornmeal, whole ground	10	5
11. Wheat bran, crude	9	5

Except for soybean flour, all of these flours are deficient in isoleucine and lysine and should be complemented by other protein sources. Legumes are the ideal match for grains since they are high in these amino acids. Brewer's yeast is also an excellent complement.

Remember, it is not only important to get enough protein, but to get it in the right proportions. The safest way to assure this is to eat different sources of protein.

Vegetables

Vegetables are not going to make a large contribution to your protein needs, but they contain many vitamins and minerals essential for good health and vitality. Most of the vegetables listed here are low in calories so you need not be concerned about eating too many of them.

Vegetables high in vitamins and minerals include: snap beans, beets, burdock, cabbage, eggplant, lettuce, onions, green peppers, pumpkins, radishes, rhubarb, squash, sweet potatoes, tomatoes, and turnips.

Many of the vegetables listed as sources of protein are deficient in sulfur content and isoleucine and should be complemented by sesame seeds and Brazil nuts. Millet, parboiled rice, and mushrooms are also high in sulfur content.

The average serving of vegetables here is based on a fresh, uncooked weight of 3½ ounces.

*Average Serving, 3½ oz.**	*Total Protein*	*Usable Protein*
1. Lima beans, green	8 grams	4 grams
2. Soybean sprouts	6	3
3. Peas, green, shelled	6	3
4. Brussels sprouts	5	3
5. Corn, 1 med. ear	4	3
6. Broccoli, 1 stalk	4	2–3
7. Kale, stems, cooked	4	2
8. Collards, cooked	4	2
9. Mushrooms	3	2
10. Asparagus	3	1.8
11. Artichoke	3	1.8
12. Cauliflower	3	1.8
13. Spinach	3	1.5
14. Turnip greens, cooked	3	1.4
15. Mung bean sprouts	4	1.4
16. Mustard greens	3	1.4
17. Potato, white, baked	2	1.2
18. Okra	2	1.2
19. Chard	2	1

*Average serving: based on fresh, uncooked weight.

Dried Fruits

The most important value of dried fruits is their content of carbohydrates and mineral matter. In order to maintain optimum health, one must have fats and fluid substances, vitamins, and carbohydrates.

Carbohydrates form the bulk of the food most people eat: breads, potatoes, vegetables, fruits, and some nuts. There are three types of carbohydrates—sugar, starch, and cellulose (gums and pectins).

The sugar content of dried fruit is far greater than of fresh fruit, but you should eat both. Fruits contain vitamins, juices, and acids. Dried fruits are rich in minerals, and only seafoods such as clams, lobsters, and oysters can compare with them. The minerals found in dried fruits are potassium, sodium, calcium, magnesium, iron, sulphur, silicone, and chlorine.

See Table page .

Nutritional Additives

The main purpose of nutritional additives is to boost your protein intake. A small amount of powdered egg white or "Tiger's Milk" can give your diet a substantial protein boost. One teaspoon of the former or ¼ cup of the latter can fill 12 to 25 percent of your daily protein requirement.

NUTRITIONAL ADDITIVES

Average Serving	*Total Protein*	*Usable Protein*
1. Egg white, powder, ½ oz.	11 grams	9 grams
2. Tiger's Milk, ¼ cup	7	5
3. Brewer's yeast, powder, 1 level tbsp.	4	2
4. Wheat germ, commercial, 2 level tbsp.	3	2

Except for Tiger's Milk, these additives are all especially high in lysine, the amino acid that is most often missing from plant sources of protein.

VITAL ELEMENTS IN COMMON FOODS

There are, of course, many other vital elements that your body needs besides proteins.

One of the important vitamins a vegetarian lacks in his diet is B_{12}, which others normally get from meat. Adelle Davis recommended that vegetarians supplement their diets by taking 50-microgram B_{12} tablets once a week to prevent a deficiency in that vitamin.

The following is a listing of the vital elements found in common foods:

CALCIUM
- Cheese (American)
- Milk (whole)
- Cheese (cottage)
- Milk (butter)
- Cauliflower
- Broccoli
- Endive
- Celery
- Beans
- Rutabagas
- Spinach
- Turnips
- Carrots
- Molasses
- Oysters
- String beans
- Cabbage
- Lettuce
- Eggs
- Nuts
- Citrus fruits
- Maple syrup
- Dried fruits
- Dried beans
- Bananas
- Pineapple
- Grapes
- Lemons

FLUORINE
- Cauliflower
- Cod liver oil
- Goat's milk
- Egg Yolk
- Cheese
- Brussels sprouts
- Milk
- Garlic
- Sauerkraut
- Seafood
- Rye Bread
- Cabbage
- Whole Grains
- Spinach
- Watercress
- Beets

HYDROGEN
- Vegetables
- Fruits
- Milk
- Water

IODINE
- Lobster
- Clams
- Oysters
- Shrimp
- Bluefish
- Mackerel
- Haddock
- Cod
- Scallops
- Halibut

CARBON
- Potatoes
- Brown Sugar
- Whole Wheat
- Shredded Wheat
- Honey

CHLORINE
- Oysters
- Cheese
- Lettuce
- Whey
- Cabbage
- Parsnips
- Beets

Turnips
Milk
Watercress
Fish
Celery
Cottage cheese
Dates
Dandelion
Coconut
Carrots
Tomatoes
Salmon
Squash
Radishes
Asparagus
Lettuce
Milk
Cabbage
Cucumber
String Beans
Spinach
Beets
Potato
Kelp
Sea Lettuce

IRON

Beans
Egg Yolk
Peas
Wheat
Oatmeal
Prunes
Spinach
Parsley
Kale
Cheese
Potato
Chard
Watercress
Oysters
Dates
Raisins
Beets
Figs
Oranges
Mushrooms
Turnips
Tomatoes
Bananas
Carrots

MAGNESIUM

Cocoa
Chocolate
Almonds
Cashews
Peanuts
Lima Beans
Whole Wheat
Brown Rice
Oatmeal
Dates
Raisins
Chard
Spinach

MANGANESE

Pineapple
Wheat
Navy Beans
Blueberries
Walnuts
Kidney Beans
Beets
Lima Beans
Gooseberries
Spinach
Peaches, dry
Blackberries
Apples
Apricots
Beet greens
Cabbage
Wheat Bran

Bananas
Watercress
Tomatoes
Apricots
Peaches
Onions
Lima Beans
Milk
Prunes
Pears
String Beans
Eggplant
Celery
Raisins
Cauliflower
Potatoes
Parsley
Citrus Fruits
Carrots

SILICON

Asparagus
Spinach
Lettuce
Barley
Figs
Berries
Oatmeal
Bran
Grapes
Strawberries
Cherries
Apples
Celery
Beets
Parsnips
Black Figs
Clams
Eggs
Nuts
Fish
Oats
Carrots

NITROGEN

Peas
Lentils
Mushrooms
Cheese
Nuts
Fish
Lima Beans

PHOSPHORUS

Lima Beans
American Cheese
Oatmeal
Fish
Eggs
Spinach
Buttermilk
Milk
Almonds
Grapes
Lentils
Pecans
Brown Rice
Walnuts
Whole Wheat
Brussels Sprouts
Corn
Dandelion Greens
Lobster
Peas
Soybeans

POTASSIUM

Cabbage
Coconut
Figs
Radishes
Chard
Onions
Eggs
Oranges

SODIUM

- Wheat Bread
- Rye Bread
- Buttermilk
- Cream Cheese
- Codfish
- Halibut
- Mackerel
- Salmon
- Bananas
- Celery
- Dandelion Greens
- Lettuce
- Spinach
- Sweet Potato
- Milk
- American Cheese
- Beets
- Watercress

SULPHUR

- Watercress
- Asparagus
- Cabbage
- Garlic
- Grapes
- Onions
- Beans
- Bran
- Bread
- Brussels Sprouts
- Cauliflower
- Cheese
- Oysters
- Peas
- Chard
- Wheat

FATS AND OILS

Fats and oils are very important in a diet. One of their best sources in a vegetarian diet is nuts. The fat content of nuts is very high, sometimes well over 50 percent.

All fats are made up of three elements—carbon, hydrogen, and oxygen. These elements are the most concentrated forms of fuel for our bodies to burn. They give twice as much heat as either proteins or carbohydrates. During digestion fats are absorbed partly into the small intestine and partly into the lymph. They ultimately go into the bloodstream.

Storage of fats takes place in deposits under the skin in the tissues of most organs, in the mesentery (the membranes of the peritoneum), and around the kidneys. When these fat storage deposits are filled, new fats pass into the liver.

You may cut down your intake of fats, but remember: you have a definite need for fat in your diet. In the form of phospholipids, fats play an important role in nourishing the brain. Fats are essential in forming acetate, a crucial element in the energy-producing cycles of your body.

Sterols, essential to life, are a class of compounds derived from fats.

The fats and oils from nuts do not contain, as far as we can tell, any Vitamin D. But margarines made from nut oils can be fortified with this important vitamin, so that they are as beneficial in this respect as butter.

Though it was once thought nut butters and oils contained no vitamin A, modern researchers have found some nut oils in fact contain substantial quantities. Since nuts contain a considerable amount of basic amino acids, butter made from nuts has a high biological value. When eaten by nursing and pregnant mothers, nut oils also improve the quality of their milk.

The digestibility of nuts increases as much as 10 percent when they are turned into butter, since the digestive juices of our bodies can work only imperfectly on nuts not ground by the teeth and broken down (they are passed undigested into the alimentary canal). Unfortunately, most commerical butters are made from roasted nuts which have been heavily salted. The heat involved during this process develops free fatty acids and destroys the B-complex vitamins. Salt and nuts interfere with their digestibility.

As one of the most concentrated foods, nuts have as high a nutritional value as any animal product (almonds and cashews taken together) except cheese. The fat in all nut butters is more easily digested than dairy butter and, unlike animal fats, mixes with water and forms an easily digested emulsion.

The digestion of fats requires the presence of organic sodium in your food. Sodium is principally an alkaline element in the process of saponification, which takes place when the pancreatic juice, bile, and intestinal juices come into contact with fats. If you eat fresh vegetables and fruits with nuts, the organic sodium found in these foods will accomplish saponification.

Much of a nut's vitamins are found in its skin, so leave this on when eating them. When nuts are salted and roasted, these vitamins are destroyed.

In conclusion, a vegetarian diet need not be low in protein. If the rules relating to protein intake, digestion, and assimilation are followed, the vegetarian should have no worry about an adequate diet.

Now we look at the combinations in which foods can be eaten to assure the body of their maximum benefit.

FOOD VALUES

Where no vitamin values are given in this table it is because they have not yet been determined. In any case, the amount present probably is insignificant for ordinary nutritional purposes, except in the case of vitamin B2.

UNIT EQUIVALENTS:

Vitamin A
1 International unit = 2 Sherman Units
= 0.6 microgram (gamma, y) of B carotene

Vitamin B_1 (Thiamine Chloride)
1 International unit = 3 micrograms (gamma, y)
= 0.003 milligram
= 2 Sherman units

Vitamin C (Ascorbic Acid, Cevitamin Acid)
1 milligram = 20 International units
= 2 Sherman units

Vitamin B_2 (Riboflavin)
1 milligram = 333 Sherman-Bourquin units
= 1000 micrograms (gamma, y)

BREADS & CEREALS	MEASURE Ordinary	Ounces	Vitamin A I.U.	Vitamin B_1 I.U.	Vitamin C Mill.	Vitamin B_2 Mill.
Barley, whole grain	1 tblsp.	½	0	3.3	—	0.001
" , pearled	1 tblsp.	½	0	0	—	
Biscuits, baking powder	1 biscuit		19	3	—	
Bread, Boston brown	1 slice 3" diam. ⅜" thick	¾	55	13	—	
" , white, made with milk	1 slice 3"x3½"x½"	¾	10	3.8	—	0.0121
" , white, made with water	1 slice 3"x3½"x½"	¾	10	3.7	—	0.006
" , 100% wheat	1 slice 3"x3"x⅜"	1	70	27	—	
" , rye	1 slice 3"x3"x⅜"	1		14	—	
Corn, whole grain yellow	1 cup	5	1,200	72	11	0.05
" , flakes, cereal	1 cup	1½		trace	0	
" , meal, white	1 cup	5	0	143		
" , meal, yellow	1 cup	5	1,200	110		
Crackers, Graham	1 cracker		26	8	0	
Custard, baked	¾ cup	3	650	21	1	

	MEASURE Ordinary	Ounces	Vitamin A I.U.	Vitamin B_1 I.U.	Vitamin C Mill.	Vitamin B_2 Mill.
Flour, rye	1 cup	5		30–70		
" , 100% whole, unbleached	1 cup	4	500	180	0	
" , white, bleached	1 cup	4	150	33	0	
" , white, pastry	1 cup	4	130	19	0	
" , white, plus germ	1 cup	4		49	0	
Griddlecakes	1 medium	2	200	11	0.5	
Muffins, plain w. egg	1 muffin		135	9	0	
" , plain, without egg	1 muffin		48	8	0	
" , bran, with egg	1 muffin		260	50	0	
" , bran, without egg	1 muffin		170	52	0	
Oatmeal, whole grain	½ cup	1¾	0	165		
" , quick cooking	½ cup	1¾	0	130	0	
Oats, rolled, packaged	½ cup	1¾	0	121	0	
" , rolled, cooked	½ cup	1¾	0	121	0	
Rice, brown	2 tblsp.	1	17	15		0.02
" , polished	2 tblsp.	1	0	0		0.022
Rolls	1 roll	1½	74	9	0	
Rye, whole grain	1 cup	5		210	0	
Wheat, whole grain	1 tblsp.	½	50	23	4	0.02
" , whole grain, cooked	1 tblsp.	½	50	23		
" , bran	1 tblsp.	½	85	28	0	

" , farina, light	1 tblsp.	½	0	0.2	0	0
" , germ	1 tblsp.	½	90	80	0	0.10
BREADS & CEREALS						
Wheat, puffed	½ cup	½		0	0	
" , semolina	1 tblsp.	½	30	7		
" , shredded	1 biscuit	1	4	20	0	0.10
Wheat, stone-ground	1 tblsp.	½	40	22	3	0.015
DAIRY PRODUCTS						
Butter	1 square	½	315	0	trace	0.001
Buttermilk	1 large glass	8	0	35	2	0.310
Cheese, Am. Cheddar	1″ cube	¾	420	3	0	0.12
" , Camembert	1″ cube	¾	750		0	
" , cottage, skim	1 tblsp.	¾	70		0	0.068
" , creamed, soft	1 tblsp.	½	310	0.6	0	0.017
" , creamed, full	piece 2″x1″	1	500		0	0.02
" , Edam	1″ cube	¾	300		0	
" , pimento (Kraft)	1″ cube	¾	500		0	
" , Roquefort	1″ cube	¾	850		0	
" , Swiss (Kraft)	piece 4″x4″	¾	440		0	
Cream, 20% fat	1 tblsp.	⅗	64	2	trace	
Cream, 40% fat	1 tblsp.	⅗	132	1.6	trace	

	MEASURE Ordinary	Ounces	Vitamin A I.U.	Vitamin B_1 I.U.	Vitamin C Mill.	Vitamin B_2 Mill.
Eggs, whole	1 egg	1½	900	19	0	0.25
" , white	1 white	9/10	0	trace	0	0.14
" , yolk	1 yolk	6/10	900	19	0	0.11
" , soft-boiled	1 egg	6/10	900	19	0	
" , hard-boiled	1 egg	6/10	900	19	0	
Milk, whole, fresh, raw	1 quart	32	1,400	150	19	1.20
	1 glass	6	260	28	3.5	0.23
" , whole, fresh, pasteur	1 glass	6	260	22	3.0	0.23
" , dried, reconstituted	1 tumbler	6	260	20	2.0	0.20
" , evaporated	½ cup	4	190	15	3.5	0.35
" , condensed, sweetened	1 tblsp.	¾	45	10		
" , skim, fresh	1 glass	6	trace	25	4.0	0.17
" , skim, dried powder	1 tblsp.	¼	0	8	0.7	0.09
" , shake, ice cream		12	0	37	4	
MEAT & FISH						
Bacon, fried	5 slices	½		5	0	
Beef, lean, top round	¼ lb.	4	40	45	2.2	0.06
Chicken, light meat	¼ lb.	4		30	0	0.029
Chicken, dark meat	¼ lb.	4		42	0	0.078

Chicken liver	⅛ lb.	2	17,000	50	11	2.0
Cod, steak, fresh	¼ lb.	4	2	34	0	0.11
Crab	¼ lb.	4	2,200	45	5	0.40
MEAT & FISH						
Halibut, muscle	¼ lb.	4		32		0.21
Ham, smoked, lean	¼ lb.	4		540		
Herring, whole	¼ lb.	4	1,700	20		0.12
Kidney, beef or calf	¼ lb.	4	450	105	12	1.6
Lamb, chop, lean	¼ lb.	4	trace	90	2	
Liver, beef, fresh	¼ lb.	4	46,000	100	34	3.4
Mackerel	¼ lb.	4		34		
Mutton, lean	¼ lb.	4		68		
Oysters, raw	⅓ cup	3½	420	75	3	0.46
Pork chop, lean	¼ lb.	4	0	515	2	
Pork, loin, lean	¼ lb.	4	0	515	2	0.28
Prawns, boiled	¼ lb.	4	1,100	20	0	0.11
Salmon, fresh, canned	¼ lb.	4	340	trace		0.27
Sardines, canned in oil	⅛ lb.	2	200	17		
Sweetbreads, fresh	¼ lb.	4		120		
Tongue, beef or sheep	¼ lb.	4		35		
Trout, fresh-water	¼ lb.	4		33		

	MEASURE					
	Ordinary	*Ounces*	*Vitamin A I.U.*	*Vitamin B_1 I.U.*	*Vitamin C Mill.*	*Vitamin B_2 Mill.*
Whiting, Atlantic	¼ lb.	4	400			
Veal, muscle, cooked	¼ lb.	4	40	45	2	0.17
FRUITS						
Apples, raw, average	1 medium	6	50	8	10	0.06
" , applesauce	½ cup	4½			9	
" , Juice	1 tumbler	6			5	
Apricots, fresh	1 medium	1⅓	1,000	3.5	2	0.04
" , dried	4 halves	⅗	540	8	0	0.009
" , dried, sulphured	4 halves	⅗	540	3	2	0.005
Avocado, California	½ avocado	4	64	40	25	0.19
Bananas	1 medium	5½	230	28	7	0.07
Blackberries	1 cup	5¾	230	10	15	
Blueberries	1 cup	5¼	150	21	8	0.021
Cantaloupe	½ melon	13½	770	73	60	0.25
Cherries, fresh, bing	10 cherries	2⅓	75	12	13	
Cranberries, fresh	1 cup	3¾	20	0	20	0
Cranberries, Juice	½ cup	2			7	0
Currant, black fresh	1 cup	3¾		11	220	
Currant, red, fresh	1 cup	3¾		16	55	
Dates, fresh	4 dates	1½	35	12	0	0.012

Dates, dried	4 dates	1½	15	8.5	0	0
Figs, fresh	3 small	4	90	30	2.3	0.06
Figs, dried	3 small	2⅕	32	30	0	
Gooseberries, canned	1 cup	4	330	56	34	
Grapes, white, seedless	½ bunch	4	22	11	2.2	0
Guava	1 guava	¾	22	2	30	0.006
Lemon	1 large	3⅗	0		55	
Lemon, Juice	2 tblsp.	1	0		15	0.001
Lime juice, fresh	2 tblsp.	1	8		10	
Mangoes	1 mango	3	850	17	25	0.05
Olives, green	6 olives	2	85	trace	0	
Olives, mission	6 olives	2	28	trace	0	
Orange, pulp	1 large	9½	80–270	95	80	0.24
Orange, Juice, fresh	1 tumbler	6	540	59	98	0.012
Orange, Juice, canned	1 tumbler	6	540	59	72	
Papayas	1 papaya	3	1,800	7	46	0.015
Peaches, fresh	1 medium	4	900	28	3.4	0.008
Peaches, canned	2 halves	3	900	22	1.7	
Pears, fresh, Bartlett	1 pear	3⅕	6	5	3	0.015
Pineapple, canned	1 slice ⅜"	1⅜	90	13	2.6	0.006
Pineapple, Canned	1 slice ⅜"	1⅜	90	7	2.3	
Pineapple, Juice, canned	1 tumbler	6	90	35	12	
Plums, fresh	2 plums	3⅘	210	17	5.5	0.027

	MEASURE Ordinary	Ounces	Vitamin A I.U.	Vitamin B_1 I.U.	Vitamin C Mill.	Vitamin B_2 Mill.
Plums, canned	2 plums	3⅘	200	15	5	
Pomegranate	1 pomegranate	6	0		20	
Prunes	4 prunes	1⅖	350	24	trace	0.3
Quince	1 quince	4		11		
Raisins	¼ cup	1	28	21	0	0.04
Raspberries	1 cup	4⅗	270	trace	33	
Strawberries	½ box	6	340	trace	93	
Tangerines	1 tangerine	2	280	22	28	0.022
Watermelon	1 slice 6¾"	11	0	25	15	0.18

VEGETABLES

	MEASURE Ordinary	Ounces	Vitamin A I.U.	Vitamin B_1 I.U.	Vitamin C Mill.	Vitamin B_2 Mill.
Artichokes	1 artichoke	8½	250	70	15	0.025
Asparagus, green	2 stalks	1⅓	130	22	19	
Beans, baked	1 cup	4	15	150	0	
" , kidney, fresh	½ cup	3	27	63	0	0.025
" , navy, dried	½ cup	3½	27	128	0	
" , lima, fresh	½ cup	3	0	100	13	0.25
" , string or snap	½ cup	3½	600	25	8	0.08
" , string, canned	½ cup	3½	600	11	2	
" , runner, green	½ cup	3½	700	40	5	
" , soy black	½ cup	3½	900	100	46	

" , soy, white	½ cup	3½	140	250		
" , wax, butter, yellow	½ cup	3½	410	30	16	
Beets, Root	½ cup diced	5	70	24	5	
VEGETABLES						
Broccoli, fresh	1 cup	4	4,000	38	80	
Brussels sprouts	6 sprouts	4	1,100	65	100	
Cabbage, white, raw	1 cup	3⅓	800	25	25	0.04
Cabbage, white, cooked	1 cup	3½	50	20	12	
Carrots, raw	1 cup diced	4¾	2,700	32	8	0.03
Carrots, cooked	1 cup diced	4¾	2,700	32	5	
Carrots, canned, strained	1 cup diced	4¾	2,700	13	3	
Cauliflower, raw	¼ head	3	49	46	33	0.065
Cauliflower, cooked	¼ head	3		35	33	
Celery, green stems	2 stalks 7"	1½	320		2	
Celery, blanched stems	2 stalks 7"	1½	2	5	2	
Chard (beet tops) raw	½ cup	5⅕	12,000	20	43	
Chard beet cooked	½ cup	5⅕	12,000	15	28	
Chicory (escarole) raw	4 leaves	3½	9,500		5	0.022
Chives	1 teasp.	⅕			2	
Collards, fresh raw	½ cup	3⅗	2,200	68	50	0.3
Collards, cooked	½ cup	3⅗	2,200	50	22	
Corn, yellow, whole	¼ cup	1⅘	250	25	5	0.018

	MEASURE Ordinary	Ounces	Vitamin A I.U.	Vitamin B_1 I.U.	Vitamin C Mill.	Vitamin B_2 Mill.
Corn, sweet, canned	½ cup	3½	120	30	11	
Cucumbers, raw	½ of 10″	6½	0–20	55	18	0.003
Dandelion greens	1 cup	3½	12,500			
Eggplant	3 slices 4″	6⅓	70	43	8	0.10
Endive	3 stalks 6″	7⅓	4,200	70	20	
Garlic	1 clove	¼	0		2	
Kale, raw	1 cup	3⅖	9,000	59	140	0.50
Kale, cooked	1 cup	3⅖	9,000	45	28	
Kohlrabi	1 cup ½″	5			39	
Leek	1 leek 7″	1	17	8	7	
Lentils	2 tblsp.	1	25	14		0.022
Lettuce, Romaine	2 leaves 9″	¾	1,300	4	1	0.03
Lettuce, Iceberg, head	¼ l. head	2½	150	20	4.5	0.05
Mustard greens	½ cup	2⅓	100	30		
Okra	5 pods	2	85	23	6	0.26
Onions, fresh	1 medium	2	6,200			
Onions, fresh, stewed	1 medium	2	0	10		
Parsley	4 stems	½	1,000		11	
Parsnips	17″x2″	6¾	380	66	38	
Peas, fresh	½ cup	2½	530	85	20	0.026
Peppers, green	1 pepper	2½	600	69	0.12	

Peppers, red	1 pepper	2½	2,200		135	
Potatoes, yellow, sweet	½ medium	3⅗	8,000	31	20	0.02
Potatoes, yellow, sweet, cooked	½ medium	3⅗	4,000	25	10	
Potatoes, white	1 medium	5⅓	60	93	30	
Potatoes, white cooked	1 medium	5⅓	40	74	30	
Rutabagas	1 cup	5	10	28	22	0.13
Sauerkraut, fresh	1 cup	5		trace	16	
Sauerkraut, cooked	1 cup	5			3	
Sauerkraut, juice	2 tblsp.	1			4	
Spinach, raw	½ cup chop	3⅗	14,000	37	53	0.065
Spinach, cooked	½ cup chop	3⅗	14,000	30	26	
Squash, winter, Hubbard	½ cup	4	2,900	21	3.5	
Squash, summer, winter	½ cup	4	170	16	3.5	
Tapioca	2 tblsp.	1	0	0	0	0
Tomatoes	1 medium	7	6,000	52	50	0.10
Tomatoes, canned	½ cup	4¼	3,600	25	24	
Tomatoes, juice, canned	½ cup	6	5,300	25	25	
Turnip greens	1 tumbler	3	6,000	38	42	0.25
Turnip, cooked	½ cup	3	6,000	25	10	
Watercress	½ bunch	1½	2,000	28	22	
NUTS						
Almonds	10 nuts	⅗	0	7	0.5	
Cashews	5 nuts	½	19			0.028

	MEASURE Ordinary	Ounces	Vitamin A I.U.	Vitamin B_1 I.U.	Vitamin C Mill.	Vitamin B_2 Mill.
Chestnuts	2 nuts	½		12	5	
Coconut, shredded	1 tblsp.	⅓	0	1	0.5	0.014
Cocoanut, milk, fresh	1 cup	8	0	0	4	
Hazel	10 nuts	½		28	2	
Peanuts, whole, Spanish	⅓ cup	1½	17	146	4	0.3
Peanuts, shelled	⅓ cup	1½	17	146	4	0.3
Peanuts, roasted	⅓ cup	1½		33	0	
Peanut butter	2 tblsp.	1½	17	125		
Pecans	12 nuts	½	14			
Pistachio nuts	15 nuts	½	14			
Walnuts	12 nuts	½	190	16	3.5	
MISCELLANEOUS						
Corn oil			0	0	0	0
Cottonseed oil			0	0	0	0
Ice cream, made with skim milk		8	0	15	3	
Lard			0	0	0	0
Mayonnaise	1 tblsp.		30	0.5	0	
Pie, apple	1 serving		90	14	0	
Pie, blueberry	1 serving		62	9	3	
Pie, chocolate	1 serving		440	10	0.5	

Soup, black bean	1 cup	4	70	11	0	
Soup, split pea	1 cup	4	145	18	0	
MISCELLANEOUS						
Soup, tomato	1 cup	4	1,480	10	9	
Soup, vegetable	1 cup	4	276	9	4	
Sugar			0	0	0	0
Tea			0	0	0	0
Coffee			0	0	0	0
Cocoa	2 tblsp.	½	0	4	0	0
Beer	1 glass	6	0	4	0	0
Yeast, baker's compressed	1 cake	2	0	100	1	1.4
Yeast, baker's dried	1 cake	½	0	70		0.5
Yeast, brewer's, fresh	1 cake	2	100	220		0.85
Yeast, brewer's, dried	1 cake	½		8–500		0.4

CHAPTER FOUR

FASTING

In this particular chapter, it is so important to the overall concepts of proper physical hygiene and well being, that we have tried to present a moderate view of the therapeutic benefits of fasting for individuals who are fasting not necessarily to rid themselves of any particular disease or metabolic condition, but rather to rest the vital digestive organs, to the small degree it does, during a limited fast, of one to five days. We have also chosen to use a question and answer format for this chapter rather than a straight narrative. This is primarily because so many people seem to be confused by the information concerning fasting, since each book on fasting seems to take a different point of view both as to the relevancy and the methods by which a person should fast. It is, therefore, easier to understand a point if it is made in the question and answer form.

We have found that most individuals who are going on a meatless or partially meatless diet also feel that some systematic fasting is necessary to help rid the body of those accumulated toxins which have resulted from a heavy meat concentrated diet. It is, therefore, relevant that this information, abbreviated as it is, supplement the other information in this text.

Q. What is a fast?

A. Fasting, which is practiced by thousands of people interested in attaining good health, is one of the most misunderstood therapeutic practices in Western society. Most people have little or no idea of the purpose of a fast. Fasting is a form of therapy one practices with his body. Fasting is a total abstinence from food for any length of time and for any variety of reasons. There are milk fasts, vegetable fasts, fruit fasts, water fasts, etc. The most common and often practiced form is total abstinence from all foods, but not water.

The body is constantly working to digest foods and to eliminate wastes, to fight disease and illness, to repair worn-out cells and to nourish the blood. By fasting with water, the body uses as little energy as possible to assimilate what it needs and to expel the remainder. Therefore, water fasting enables the body to rest completely.

Fasting is not a cure for any disease or ailment. It allows the body full range and scope to fulfill its self-healing, self-repairing, self-rejuvenating functions to the best advantage. Healing is an internal biological process. Fasting allows your system a physiological rest and permits the organism to become 100 percent efficient in healing itself.

Fasting gives the overworked internal organs and tissues rest and time for rehabilitation. Fasting exhilarates the internal power and vitality of your system to flush out toxic matter and poisons which have accumulated over the years. Fasting improves circulation and promotes vigor, endurance, stamina, and strength. Fasting renovates, revivifies and purifies each one of the millions of cells which make up your body.

Q. Does fasting help cleanse the body?

A. Although there has not yet been controlled clinical testing of this theory, it is known that the body accumulates thousands of chemical substances in the course of normal living. Tests have shown that certain toxins, including DDT, mercury, numerous food additives, and air and water pollutants become stored in the system, usually in the fat cells, but also in feces carrying waste products, which can become impacted and remain along the colon walls for long periods of time, producing toxic effects.

Since most toxic substances accumulate in the body's fat cells, short-term fasts (five–six days) tend to have little results, for loss of water and protein (from lean cells) occurs during this time, before the body begins to break down fat cells. Fasts of longer duration, however, which do result in the breakdown of fat cells, can release many unwanted substances from the system.

Q. Many claim that fasting gives the body a chance to rest and accomplish needed repairs, is this true?

A. During periods of illness, most animals and humans tend to refrain from food. What causes this tendency is still a mystery. What is known, however, is that the body uses considerable energy in the digestion, absorption, and elimination of food. Research has shown that bodily expenditure while fasting is decreased by approximately 20 percent on bed rest and more than 30 percent if ambulation occurs. It is clear that by refraining from food the body conserves energy which can be vital in the struggle against illness and disease.

Q. Should one take vitamin supplements during a fast?

A. With some exceptions, a person on a well-balanced diet would normally have a good reserve of essential vitamins and minerals, especially the fat soluble vitamins A, D, and E. But clinical tests have shown that most water dispersible vitamins (B complex and C), and certain minerals (including potassium, calcium, magnesium, sodium and chloride), the reserves of which are never very high, tend to be eliminated in large amounts during the first week of fasting. After this time, they tend to decrease, being expelled more slowly and at a more even rate until the body adjusts to the fast.

Specific loss rates have been estimated as follows: it takes only four to ten days for thiamin depletion to be detected; this can be corrected by administering 50 mg. of B_1 followed by daily doses of 1 mg. Pantothenic acid is excreted very rapidly for the first ten days, then the rate of decrease declines. Unlike pantothenic acid, biotin continues to be excreted at the same rate, even after twenty-five days of fasting. Riboflavin (B_2) falls to a very low level after ten days, to 30 percent of the loss during the first ten days. Pyrodoxine (B_6) loss diminishes as the duration of the fast lengthens. Potassium depletion is high during the first five days of a fast, then declines and maintains a stable rate. The same is true of sodium and chlorine. One report suggested supplementation of vitamin E along with A and D should be undertaken by a person on a fast because a decrease in tocopherol level is in direct parallel to weight loss. This report also showed that 50 percent of the body's tocopherol was reduced after two weeks of fasting.

In view of these findings, it is suggested that vitamins and minerals, especially the B complex and potassium, should be supplemented during a fast of more than one week's duration.

Q. Can fasting be dangerous?

A. Most people tolerate fasting well, but a few test cases have shown that it may not be advisable under certain conditions, specifically for persons with a history of heart, liver, or kidney disorders. If one is reasonably healthy and proceeds under trained supervision, there is little chance that fasting will prove dangerous.

Q. What are the side effects of fasting?

A. The body requires about two to three days to burn up its glycogen stores and then it must depend upon triglycerides mobilized from adipose tissue for energy. This results in a keto acid rise in the serum, which is noticeable at once in a foul taste in the mouth and a sour breath, which lasts for the duration of the fast. This effect can be offset with a strong mouthwash.

During the early stages of fasting as much protein is lost as fat, and this loss may result in arrest of hair growth and/or dry and scaly skin. Some cases have shown a loss of hair during the fast, but after the fast hair growth returned to normal. There is a case of a man who, after forty years of baldness, regained some hair growth following a fast. Others, with a history of sparse hair, notice an increase in hair after fasting. Fasting apparently influences the hormone system, and thus it could possibly correct problems which are a result of hormone disturbance or imbalance. In addition to affecting hair growth, fasts have regulated the menstrual cycles of women with a history of late periods.

Sebum production, which is also regulated by hormones, tends to drop about 40 percent after fasts of two weeks or more, and this results in a drying of the skin. This effect can be mitigated by the use of body creams and lotions; unsaturated vegetable oil may also be applied.

During the first three days of fasting, a person may experience nausea, headaches, dizziness, or fatigue, but these effects usually cease after three days. Long-term effects of fasting may include mild anemia, which may be offset by supplementation of iron and B_{12}.

From a psychological viewpoint, fasting has shown no significant impairment of personality or faculties, and in some cases has even caused some improvement. Whereas some persons may experience periods of irritability, mild depression or feelings of dependency, in others fasting has been shown to produce euphoric feelings, and increased facility in areas of learning, mental control, and memory.

Q. How can one curb hunger while fasting?

A. Hunger is greatest during the first three days of a fast, after which it sharply declines, so that four or five days after having begun a fast feelings of hunger are either hardly perceptible or not felt at all. An English physician, Dr. T. J. Tomson, reports having had obese patients on fasts for up to 249 days with no signs of hunger. It is advised, therefore, that for the first three days one simply drink water and try to keep mentally occupied until feelings of hunger diminish and disappear.

Q. Should one exercise during fasts?

A. No. While fasting, one should get as much rest as possible and limit all physical activity. Leisurely walking is the best exercise, and is well tolerated.

Q. How much water should one drink while fasting?

A. Two quarts a day are suggested.

Q. Can fasting reduce the cholesterol level?

A. Serum cholesterol levels usually drop during fasting. but after refeeding rise again to the previous level.

Q. Should one take diuretics or laxatives while fasting?

A. During the early days of a fast the body loses large amounts of potassium, following which the rate of loss decreases and stabilizes. Since diuretics result in potassium loss in most cases, they are not advisable for persons on fasts.

Prior to fasting, one should eat quantities of roughage foods like celery, lettuce, grainy bread, carrots, etc. to provide bulk for elimination. Then, laxatives should only be taken during the first few days of a fast. Peristalsis ceases after the early days of a fast and periodic enemas are needed to remove impacted foods lining the colon walls.

Q. Is fasting a good way to lose weight?

A. Under certain circumstances fasting has proven to be very beneficial in weight lose. The weight of the person is of prime importance because obese people can tolerate a prolonged fast better than lighter people. During the early stages of fasting (first four days–fourteen days) weight loss is substantial, but it is primarily water loss and easily gained back. Weight loss is greatest in heavier people than with lighter people. Also during the early stages as much protein is lost as fat which is not desirable. Fat people surprisingly have more lean meat than thinner people and can afford the loss. As the fast progresses, fat loss increases more than protein loss. The rate of weight loss is normally predictable. The first day the most is lost, then over the next ten days average two–three lbs., then a somewhat steady rate of .8 lb. a day for a month or month and a half, then a constant trend for the remainder of the fast.

Once the body depletes its glucose store (which is very little), it uses protein for energy. The amount of this fuel will depend on prior eating habits. Tests have shown that small amounts of carbohydrate taken while on fasts has the ability of sparing both water excretion (by retaining sodium) and protein usage (by 50 percent) which results in a lower rate of weight loss.

Q. How should one break a fast?

A. Milk or vegetable juice along with tea and honey are well tolerated. Refeeding should be a gradual process, working slowly toward easily digested vegetables and then eventually a return to the normal diet.

Q. When should one terminate the fast?

A. If there are any unpleasant side effects, pain, nausea, dizziness, after the first few days; then by all means stop.

After a week if there is a fall in total body potassium, discontinue the fast. Failure to develop acidosis or its sudden improvement is evidence that the fast has been broken. Sodium bicarbonate is used effectively for those people with dizziness and nausea as a result of acidosis. Any significant changes in the body's pH acid-alkaline balance after a week normally is cause for breaking the fast.

Massive fluid loss in the beginning brings about a fall in extra cellular fluid which may result in hypotension. If hypotension reaches a high degree, fasting should be stopped, especially for people with a history of ischemic diseases of the brain or heart.

Fasting can be safe if a person is aware of changes in his system and is supervised by a physician.

CHAPTER FIVE

FOOD VALUE TABLES

The following tables can be used by the reader in determining the precise food values of nearly 500 food items commonly used in the United States. The values are based on data from the U.S. Department of Agriculture. With the use of these charts it should be easy for the reader to plan his daily menu to include all the nutritional elements necessary for good health.

This information is broken down into two tables. The first covers the nutritive values of the edible part of foods. The second is a compilation of the amino acid content of foods. This table gives you the average amounts of the eight "essential" amino acids found in each food, plus arginine and histidine.

NUTRITIVE VALUES OF THE EDIBLE PART OF FOODS

Reprinted from *Nutritive Value of Foods* (U.S. Department of Agriculture, Home and Garden Bulletin No. 72).

[Dashes show that no basis could be found for imputing a value although there was some reason to believe that a measurable amount of the constituent might be present]

Food, Approximate Measure, and Weight (in Grams)		Food Energy	Protein	Fat (Total Lipid)	Fatty acids: Saturated (Total)	Fatty acids: Unsaturated: Oleic	Fatty acids: Unsaturated: Linoleic	Carbohydrate	Calcium	Iron	Vitamin A Value	Thiamine	Riboflavin	Niacin	Ascorbic Acid
MILK, CREAM, CHEESE; RELATED PRODUCTS															
Milk, cow's:	Grams	(Calories)	(Gm.)	(Gm.)	(Gm.)	(Gm.)	(Gm.)	(Gm.)	(Mg.)	(Mg.)	(I.U.)	(Mg.)	(Mg.)	(Mg.)	(Mg.)
Fluid, whole (3.5% fat). 1 cup	244	160	9	9	5	3	Trace	12	288	0.1	350	0.08	0.42	0.1	2
Fluid, nonfat (skim) 1 cup	246	90	9	Trace	——	——	——	13	298	.1	10	.10	.44	.2	2
Buttermilk, cultured, from skim milk. 1 cup	246	90	9	Trace	——	——	——	13	298	.1	10	.09	.44	.2	2
Evaporated, unsweetened, undiluted. 1 cup	252	345	18	20	11	7	1	24	635	.3	820	.10	.84	.5	3
Condensed, sweetened, undiluted. 1 cup	306	980	25	27	15	9	1	166	802	.3	1,090	.23	1.17	.5	3
Dry, whole 1 cup	103	515	27	28	16	9	1	39	936	.5	1,160	.30	1.50	.7	6
Dry, nonfat, instant 1 cup	70	250	25	Trace	——	——	——	36	905	.4	20	.24	1.25	.6	5
Milk, goat's:															
Fluid, whole 1 cup	244	165	8	10	6	2	Trace	11	315	.2	390	.10	.27	.7	2

Cream:															
Half-and-half (cream and milk).															
1 cup	242	325	8	28	16	9	1	11	261	.1	1,160	.08	.38	.1	2
1 tablespoon	15	20	Trace	2	1	1	Trace	1	16	Trace	70	Trace	.02	Trace	Trace
Light, coffee or table															
1 cup	240	505	7	49	27	16	1	10	245	.1	2,030	.07	.36	.1	2
1 tablespoon	15	30	Trace	3	2	1	Trace	1	15	Trace	130	Trace	.02	Trace	Trace
Whipping, unwhipped (volume about double when whipped):															
Light															
1 cup	239	715	6	75	41	25	2	9	203	.1	3,070	.06	.30	.1	2
1 tablespoon	15	45	Trace	5	3	2	Trace	1	13	Trace	190	Trace	.02	Trace	Trace
Heavy															
1 cup	238	840	5	89	49	29	3	7	178	.1	3,670	.05	.26	.1	2
					3	2	Trace	Trace	11	Trace	230	Trace	.02	Trace	Trace
Cheese:															
Blue or Roquefort type															
1 ounce	28	105	6	9	5	3	Trace	1	89	.1	350	.01	.17	.1	0
Cheddar or American:															
Ungrated															
1 inch cube	17	70	4	5	3	2	Trace	Trace	128	.2	220	Trace	.08	Trace	0
Grated															
1 cup	112	445	28	36	20	12	1	2	840	1.1	1,470	.03	.51	.1	0
1 tablespoon	7	30	2	2	1	1	Trace	Trace	52	.1	90	Trace	.03	Trace	0
Cheddar, process															
1 ounce	28	105	7	9	5	3	Trace	1	219	.3	350	Trace	.12	Trace	0
Cheese foods, Cheddar															
1 ounce	28	90	6	7	4	2	Trace	2	162	.2	280	.01	.16	Trace	0
Cottage cheese, from skim milk:															
Creamed															
1 cup	225	240	31	9	5	3	Trace	7	212	0.7	380	0.07	0.56	0.2	0
1 ounce	28	30	4	1	1	Trace	Trace	1	27	.1	50	.01	.07	Trace	0
Uncreamed															
1 cup	225	195	38	1	Trace	Trace	Trace	6	202	.9	20	.07	.63	.2	0
1 ounce	28	25	5	Trace	----	----	----	1	26	.1	Trace	.01	.08	Trace	0
Cream cheese															
1 ounce	28	105	2	11	6	4	Trace	1	18	.1	440	Trace	.07	Trace	0
1 tablespoon	15	55	1	6	3	2	Trace	Trace	9	Trace	230	Trace	.04	Trace	0

Food, Approximate Measure, and Weight (in Grams)		Food Energy	Protein	Fat (Total Lipid)	Fatty acids: Saturated (Total)	Unsaturated: Oleic	Unsaturated: Linoleic	Carbohydrate	Calcium	Iron	Vitamin A Value	Thiamine	Riboflavin	Niacin	Ascorbic Acid
	Grams	(Calories)	(Gm.)	(Gm.)	(Gm.)	3 (Gm.)	Trace (Gm.)	1 (Gm.)	262 (Mo.)	.3 (Mo.)	320 (I.U.)	Trace (Mo.)	.11 (Mo.)	Trace (Mo.)	0 (Mo.)
CHEESE															
Swiss (domestic) 1 ounce	28	105	8	8	4										
Milk beverages:															
Cocoa 1 cup	242	235	9	11	6	4	Trace	26	286	.9	390	.09	.45	.4	2
Chocolate-flavored milk drink (made with skim milk). 1 cup	250	190	8	6	3	2	Trace	27	270	.4	210	.09	.41	.2	2
Malted milk 1 cup	270	280	13	12	----	----	----	32	364	.8	670	.17	.56	.2	2
Milk desserts:															
Cornstarch pudding, plain (blanc mange). 1 cup	248	275	9	10	5	3	Trace	39	290	.1	390	.07	.40	.1	2
Custard, baked 1 cup	248	285	13	14	6	5	1	28	278	1.0	870	.10	.47	.2	1
Ice cream, plain, factory packed:															
Slice or cut brick, 1/8 of quart brick. 1 slice or cut brick	71	145	3	9	5	3	Trace	15	87	.1	370	.03	.13	.1	1
Container 3½ fluid ounces	62	130	2	8	4	3	Trace	13	76	.1	320	.03	.12	.1	1
Container 8 fluid ounces	142	295	6	18	10	6	1	28	175	.1	740	.06	.27	.1	1
Ice Milk 1 cup	187	285	9	10	6	3	Trace	42	292	.2	390	.09	.41	.2	2
Yogurt, from partially skimmed milk. 1 cup	246	120	8	4	2	1	Trace	13	295	.1	170	.09	.43	.2	2

EGGS															
Eggs, large, 24 ounces per dozen:															
Raw:															
Whole, without shell															
1 egg	50	80	6	6	2	3	Trace	Trace	27	1.1	590	.05	.15	Trace	0
White of egg															
1 white	33	15	4	Trace	----	----	----	Trace	3	Trace	0	Trace	.09	Trace	0
Yolk of egg															
1 yolk	17	60	3	5	2	2	Trace	Trace	24	.9	580	.04	.07	Trace	0
Cooked:															
Boiled, shell removed															
2 eggs	100	160	13	12	4	5	1	1	54	2.3	1,180	.09	.28	.1	0
Scrambled, with milk and fat.															
1 egg	64	110	7	8	3	3	Trace	1	51	1.1	690	.05	.18	Trace	0
MEAT, POULTRY, FISH, SHELLFISH; RELATED PRODUCTS															
Bacon, broiled or fried, crisp.															
2 slices	16	100	5	8	3	4	1	1	2	.5	0	.08	.05	.8	----
Beef, trimmed to retail basis,[1] cooked:															
Cuts braised, simmered, or pot-roasted:															
Lean and fat															
3 ounces	85	245	23	16	8	7	Trace	0	10	2.9	30	.04	.18	3.5	----
Lean only															
2.5 ounces	72	140	22	5	2	2	Trace	0	10	2.7	10	.04	.16	3.3	----
Hamburger (ground beef), broiled:															
Lean															
3 ounces	85	185	23	10	5	4	Trace	0	10	3.0	20	.08	.20	5.1	----
Regular															
3 ounces	85	245	21	17	8	8	Trace	0	9	2.7	30	.07	.18	4.6	----
Roast, oven-cooked, no liquid added:															
Relatively fat, such as rib:															
Lean and fat															
3 ounces	85	375	17	34	16	15	1	0	8	2.2	70	.05	.13	3.1	----
Lean only															
1.8 ounces	51	125	14	7	3	3	Trace	0	6	1.8	10	.04	.11	2.6	----

[1]Outer layer of fat on the cut was removed to withing approximately ½ inch of the lean. Deposits of fat within the cut were not removed.

Food, Approximate Measure, and Weight (in Grams)		Food Energy	Protein	Fat (Total Lipid)	Fatty acids: Saturated (Total)	Fatty acids: Unsaturated: Oleic	Fatty acids: Unsaturated: Linoleic	Carbohydrate	Calcium	Iron	Vitamin A Value	Thiamine	Riboflavin	Niacin	Ascorbic Acid
	Grams	(Calories)	(Gm.)	(Gm.)	(Gm.)	(Gm.)	(Gm.)	(Gm.)	(Mg.)	(Mg.)	(I.U.)	(Mg.)	(Mg.)	(Mg.)	(Mg.)
Relatively lean, such as heel of round:															
Lean and fat 3 ounces	85	165	25	7	3	3	Trace	0	11	3.2	10	.06	.19	4.5	----
Lean only 2.7 ounces	78	125	24	3	1	1	Trace	0	10	3.0	Trace	.06	.18	4.3	----
Steak, broiled:															
Relatively fat, such as sirloin:															
Lean and fat 3 ounces	85	330	20	27	13	12	1	0	9	2.5	50	.05	.16	4.0	----
Lean only 2.0 ounces	56	115	18	4	2	2	Trace	0	7	2.2	10	.05	.14	3.6	----
Relatively lean, such as round:															
Lean and fat 3 ounces	85	220	24	13	6	6	Trace	0	10	3.0	20	.07	.19	4.8	----
Lean only 2.4 ounces	68	130	21	4	2	2	Trace	0	9	2.5	10	.06	.16	4.1	----
Beef, canned:															
Corned beef 3 ounces	85	185	22	10	5	4	Trace	0	17	3.7	20	.01	.20	2.9	----
Corned beef hash 3 ounces	85	155	7	10	5	4	Trace	9	11	1.7	----	.01	.08	1.8	----
Beef, dried or chipped 2 ounces	57	115	19	4	2	2	Trace	0	11	2.9	----	.04	.18	2.2	----
Beef and vegetable stew 1 cup	235	210	15	10	5	4	Trace	15	28	2.8	2,310	.13	.17	4.4	15
Beef potpie, baked: Individual pie, 4¼-inch diameter, weight before															

baking about 8 ounces.															
1 pie	227	560	23	33	9	20	2	43	32	4.1	1,860	.25	.27	4.5	7
Chicken, cooked:															
Flesh only, broiled															
3 ounces	85	115	20	3	1	1	1	0	8	1.4	80	0.05	0.16	7.4	—
Breast, fried, ½ breast:															
With bone															
3.3 ounces	94	155	25	5	1	2	1	1	9	1.3	70	.04	.17	11.2	—
Flesh and skin only															
2.7 ounces	76	155	25	5	1	2	1	1	9	1.3	70	.04	.17	11.2	—
Drumstick, fried:															
With bone															
2.1 ounces	59	90	12	4	1	2	1	Trace	6	.9	50	.03	.15	2.7	—
Flesh and skin only															
1.3 ounces	38	90	12	4	1	2	1	Trace	6	.9	50	.03	.15	2.7	—
Chicken, canned, boneless															
3 ounces	85	170	18	10	3	4	2	0	18	1.3	200	.03	.11	3.7	3
Chicken potpie. See Poultry potpie.															
Chile con carne, canned:															
With beans															
1 cup	250	335	19	15	7	7	Trace	30	80	4.2	150	.08	.18	3.2	—
Without beans															
1 cup	255	510	26	38	18	17	1	15	97	3.6	380	.05	.31	5.6	—
Heart, beef, lean, braised															
3 ounces	85	160	27	5	—	—	—	1	5	5.0	20	.21	1.04	6.5	1
Lamb, trimmed to retail basis,[1] cooked:															
Chop, thick, with bone, broiled.															
1 chop, 4.8 ounces	137	400	25	33	18	12	1	0	10	1.5	—	.14	.25	5.6	—
Lean and fat															
4.0 ounces	112	400	25	33	18	12	1	0	10	1.5	—	.14	.25	5.6	—
Lean only															
2.6 ounces	74	140	21	6	3	2	Trace	0	9	1.5	—	.11	.20	4.5	—
Leg, roasted:															
Lean and fat															
3 ounces	85	235	22	16	9	6	Trace	0	9	1.4	—	.13	.23	4.7	—
Lean only															
2.5 ounces	71	130	20	5	3	2	Trace	0	9	1.4	—	.12	.21	4.4	—

Food, Approximate Measure, and Weight (in Grams)		Food Energy	Protein	Fat (Total Lipid)	Fatty acids: Saturated (Total)	Fatty acids: Unsaturated Oleic	Fatty acids: Unsaturated Linoleic	Carbohydrate	Calcium	Iron	Vitamin A Value	Thiamine	Riboflavin	Niacin	Ascorbic Acid
	Grams	(Calories)	(Gm.)	(Gm.)	(Gm.)	(Gm.)	(Gm.)	(Gm.)	(Mg.)	(Mg.)	(I.U.)	(Mg.)	(Mg.)	(Mg.)	(Mg.)
Shoulder, roasted:															
Lean and fat 3 ounces	85	285	18	23	13	8	1	0	9	1.0	——	.11	.20	4.0	——
Lean only 2.3 ounces	64	130	17	6	3	2	Trace	0	8	1.0	——	.10	.18	3.7	——
Liver, beef, fried 2 ounces	57	130	15	6	——	——	——	3	6	5.0	30,280	.15	2.37	9.4	15
Pork, cured, cooked:															
Ham, light cure, lean and fat, roasted. 3 ounces	85	245	18	19	7	8	2	0	8	2.2	0	.40	.16	3.1	——
Luncheon meat:															
Boiled ham, sliced 2 ounces	57	135	11	10	4	4	1	0	6	1.6	0	.25	.09	1.5	——
Canned, spiced or unspiced. 2 ounces	57	165	8	14	5	6	1	1	5	1.2	0	.18	.12	1.6	——
Pork, fresh, trimmed to retail basis,[1] cooked:															
Chop, thick, with bone 1 chop, 3.5 ounces	98	260	16	21	8	9	2	0	8	2.2	0	.63	.18	3.8	——
Lean and fat 2.3 ounces	66	260	16	21	8	9	2	0	8	2.2	0	.63	.18	3.8	——
Lean only 1.7 ounces	48	130	15	7	2	3	1	0	7	1.9	0	.54	.16	3.3	——
Roast, oven-cooked, no liquid added:															
Lean and fat															

[1]Outer layer of fat on the cut was removed to within approximately ½ inch of the lean. Deposits of fat within the cut were not removed.

3 ounces	85	310	21	24	9	10	2	0	9	2.7	0	.78	.22	4.7	—
Lean only															
2.4 ounces	68	175	20	10	3	4	1	0	9	2.6	0	.73	.21	4.4	—
Cuts, simmered:															
Lean and fat															
3 ounces	85	320	20	26	9	11	2	0	8	2.5	0	.46	.21	4.1	—
Lean only															
2.2 ounces	63	135	18	6	2	3	1	0	8	2.3	0	.42	.19	3.7	—
Poultry potpie (based on chicken potpie). Individual pie, 4¼-inch diameter, weight before baking, about															
1 pie	227	535	23	31	10	15	3	42	68	3.0	3,020	.25	.26	4.1	5
Sausage:															
Bologna, slice, 4.1 by 0.1 inch.															
8 slices	227	690	27	62	—	—	—	2	16	4.1	—	.36	.49	6.0	—
Frankfurter, cooked															
1 frankfurter	51	155	6	14	—	—	—	1	3	.8	—	.08	.10	1.3	—
Pork, links or patty, cooked.															
4 ounces	113	540	21	50	18	21	5	Trace	8	2.7	0	.89	.39	4.2	—
Tongue, beef, braised															
3 ounces	85	210	18	14	—	—	—	Trace	6	1.9	—	.04	.25	3.0	—
Turkey potpie. See Poultry potpie.															
Veal, cooked:															
Cutlet, without bone, broiled.															
3 ounces	85	185	23	9	5	4	Trace	—	9	2.7	—	.06	.21	4.6	—
Roast, medium fat, medium done; lean and fat.															
3 ounces	85	230	23	14	7	6	Trace	0	10	2.9	—	.11	.26	6.6	—
Fish and shellfish:															
Bluefish, baked or broiled.															
3 ounces	85	135	22	4	—	—	—	0	25	.6	40	.09	.08	1.6	—
Clams:															
Raw, meat only															
3 ounces	85	65	11	1	—	—	—	2	59	5.2	90	.08	.15	1.1	8
Canned, solids and liquid.															
3 ounces	85	45	7	1	—	—	—	2	47	3.5	—	.01	.09	.9	—
Crabmeat, canned															
3 ounces	85	85	15	2	—	—	—	1	38	.7	—	.07	.07	1.6	—

Food, Approximate Measure, and Weight (in Grams)		Food Energy	Protein	Fat (Total Lipid)	Fatty acids			Carbohydrate	Calcium	Iron	Vitamin A Value	Thiamine	Riboflavin	Niacin	Ascorbic Acid
					Saturated (Total)	Unsaturated									
						Oleic	Linoleic								
	Grams	(Calories)	(Gm.)	(Gm.)	(Gm.)	(Gm.)	(Gm.)	(Gm.)	(Mo.)	(Mo.)	(I.U.)	(Mo.)	(Mo.)	(Mo.)	(Mo.)
Fish sticks, breaded, cooked, frozen; stick, 3.8 by 1.0 by 0.5 inch. 10 sticks or 8-ounce pacage	227	400	38	20	5	4	10	15	25	.9	——	.09	.16	3.6	——
Fish and shellfish—Continued															
Haddock, fried 3 ounces	85	140	17	5	1	3		5	34	1.0	——	0.03	0.06	2.7	2
Mackerel:															
Broiled, Atlantic 3 ounces	85	200	19	13	——	——	——	0	5	1.0	450	.13	.23	6.5	——
Canned, Pacific, solids and liquid.[2]					——	——	——	0	221	1.9	20	.02	.27	7.4	——
Ocean perch, breaded (egg and breadcrumbs), fried. 3 ounces	85	195	16	11	——	——	——	6	28	1.1	——	.08	.09	1.5	——
Oysters, meat only:															
Raw, 13-19 medium selects. 1 cup	240	160	20	4	——	——	——	8	226	13.2	740	.33	.43	6.0	——
Oyster stew, 1 part oysters to 3 parts milk by volume, 3-4 oysters. 1 cup	230	200	11	12	——	——	——	11	269	3.3	640	.13	.41	1.6	——
Salmon, pink canned 3 ounces	85	120	17	5	1	1	Trace	0	167[3]	.7	60	.03	.16	6.8	——
Sardines, Atlantic, canned in oil, drained solids. 3 ounces	85	175	20	9	——	——	——	0	372	2.5	190	.02	.17	4.6	——

[2]Vitamin values based on drained solids.
[3]Based on total contents of can. If bones are discarded, value will be greatly reduced.

Food															
Shad, baked															
3 ounces	85	170	20	10	—	—	—	0	20	.5	20	.11	.22	7.3	—
Shrimp, canned, meat only.															
3 ounces	85	100	21	1	—	—	—	1	98	2.6	50	.01	.03	1.5	—
Swordfish, broiled with butter or margarine.															
3 ounces	85	150	24	5	—	—	—	0	23	1.1	1,750	.03	.04	9.3	—
Tuna, canned in oil, drained solids.															
3 ounces	85	170	24	7	—	—	—	0	7	1.6	70	.04	.10	10.1	—
Almonds, shelled															
1 cup	142	850	26	77	6	52	15	28	332	6.7	0	.34	1.31	5.0	Trace
Beans, dry:															
Common varieties, such as Great Northern, navy, and others, canned:															
Red															
1 cup	256	230	15	1	—	—	—	42	74	4.6	Trace	.13	.10	1.5	—
White, with tomato sauce:															
With pork															
1 cup	261	320	16	7	3	3	1	50	141	4.7	340	.20	.08	1.5	5
Without pork															
1 cup	261	310	16	1	—	—	—	60	177	5.2	160	.18	.09	1.5	5
Lima, cooked															
1 cup	192	260	16	1	—	—	—	48	56	5.6	Trace	.26	.12	1.3	Trace
Brazil nuts															
1 cup	140	915	20	94	19	45	24	15	260	4.8	Trace	1.34	.17	2.2	—
Cashew nuts, roasted															
1 cup	135	760	23	62	10	43	4	40	51	5.1	140	.58	.33	2.4	—
Coconut:															
Fresh, shredded															
1 cup	97	335	3	34	29	2	Trace	9	13	1.6	0	.05	.02	.5	3
Dried, shredded, sweetened.															
1 cup	62	340	2	24	21	2	Trace	33	10	1.2	0	.02	.02	.2	0
Cowpeas or blackeye peas, dry, cooked.															
1 cup	248	190	13	1	—	—	—	34	42	3.2	20	.41	.11	1.1	Trace
Peanuts, roasted, salted:															
Halves															
1 cup	144	840	37	72	16	31	21	27	107	3.0	—	.46	.19	24.7	0
Chopped															
1 tablespoon	9	55	2	1	1	2	1	2	7	.2	—	.03	.01	1.5	0

Food, Approximate Measure, and Weight (in Grams)		Food Energy	Protein	Fat (Total Lipid)	Fatty acids: Saturated (Total)	Fatty acids: Unsaturated Oleic	Fatty acids: Unsaturated Linoleic	Carbohydrate	Calcium	Iron	Vitamin A Value	Thiamine	Riboflavin	Niacin	Ascorbic Acid
	Grams	(Calories)	(Gm.)	(Gm.)	(Gm.)	(Gm.)	(Gm.)	(Gm.)	(Mo.)	(Mo.)	(I.U.)	(Mo.)	(Mo.)	(Mo.)	(Mo.)
Peanut butter															
1 tablespoon	16	95	4	8	2	4	2	3	9	.3	----	.02	.02	2.4	0
Peas, split, dry, cooked															
1 cup	250	290	20	1	----	----	----	52	28	4.2	100	.37	.22	2.2	----
Pecans:															
Halves															
1 cup	108	740	10	77	5	48	15	16	79	2.6	140	.93	.14	1.0	2
Chopped															
1 tablespoon	7.5	50	1	5	Trace	3	1	1	5	.2	10	.06	.01	.1	Trace
Walnuts, shelled:															
Black or native, chopped.															
1 cup	126	790	26	75	4	26	36	19	Trace	7.6	380	.28	.14	.9	----
English or Persian:															
Halves															
1 cup	100	650	15	64	4	10	40	16	99	3.1	30	.33	.13	.9	3
Chopped															
1 tablespoon	8	50	1	5	Trace	1	3	1	8	.2	Trace	.03	.01	.1	Trace
VEGETABLES AND VEGETABLE PRODUCTS															
Asparagus:															
Cooked, cut spears															
1 cup	175	35	4	Trace	----	----	----	6	37	1.0	1,580	.27	.32	2.4	46
Canned spears, medium:															
Green															
6 spears	96	20	2	Trace	----	----	----	3	18	1.8	770	.06	.10	.8	14
Bleached															
6 spears	96	20	2	Trace	----	----	----	4	15	1.0	80	.05	.06	.7	14

Beans:															
Lima, immature, cooked															
1 cup	160	180	12	1	—	—	—	32	75	4.0	450	.29	.16	2.0	28
Snap, green:															
Cooked:															
In small amount of water, short time.															
1 cup	125	30	2	Trace	—	—	—	7	62	.8	680	.08	.11	.6	16
In large amount of water, long time.															
1 cup	125	30	2	Trace	—	—	—	7	62	0.8	680	0.07	0.10	0.4	13
Canned:															
Solids and liquid															
Strained or chopped (baby food).															
1 cup	239	45	2	Trace	—	—	—	10	81	2.9	690	.08	.10	.7	9
1 ounce	28	5	Trace	Trace	—	—	—	1	9	.3	110	.01	.02	.1	Trace
Bean sprouts. See Sprouts.															
Beets, cooked, diced															
1 cup	165	50	2	Trace	—	—	—	12	23	.8	40	.04	.07	.5	11
Broccoli spears, cooked															
1 cup	150	40	5	Trace	—	—	—	7	132	1.2	3,750	.14	.29	1.2	135
Brussels sprouts, cooked															
1 cup	130	45	5	1											
Cabbage:															
Raw:															
Finely shredded															
1 cup	100	25	1	Trace	2	2	5	9	52	.5	180	.06	.06	.3	35
Coleslaw															
1 cup	120	120	1	9	—	—	—	5	49	.4	130	.05	.05	.3	47
Cooked:															
In small amount of water, short time.															
1 cup	170	35	2	Trace	—	—	—	7	75	.5	220	.07	.07	.5	56
In large amount of water, long time.															
1 cup	170	30	2	Trace	—	—	—	7	71	.5	200	.04	.04	.2	40

Food, Approximate Measure, and Weight (In Grams)		Food Energy	Protein	Fat (Total Lipid)	Fatty acids: Saturated (Total)	Fatty acids: Unsaturated Oleic	Fatty acids: Unsaturated Linoleic	Carbohydrate	Calcium	Iron	Vitamin A Value	Thiamine	Riboflavin	Niacin	Ascorbic Acid
	Grams	(Calories)	(Gm.)	(Gm.)	(Gm.)	(Gm.)	(Gm.)	(Gm.)	(Mo.)	(Mo.)	(I.U.)	(Mo.)	(Mo.)	(Mo.)	(Mo.)
Cabbage, celery or Chinese:															
Raw, leaves and stalk, 1-inch pieces. 1 cup	100	15	1	Trace	—	—	—	3	43	.6	150	.05	.04	.6	25
Cabbage, spoon (or pakchoy), cooked. 1 cup	150	20	2	Trace	—	—	—	4	222	.9	4,650	.07	.12	1.1	23
Carrots:															
Raw:															
Whole, 5½ by 1-inch, (25 thin strips). 1 carrot	50	20	1	Trace	—	—	—	5	18	.4	5,500	.03	.03	.3	4
Grated 1 cup	110	45	1	Trace	—	—	—	11	41	.8	12,100	.06	.06	.7	9
Cooked, diced 1 cup	145	45	1	Trace	—	—	—	10	49	.9	15,220	.08	.07	.7	9
Canned, strained or chopped (baby food). 1 ounce	28	10	Trace	Trace	—	—	—	2	7	.1	3,690	.01	.01	.1	1
Cauliflower, cooked, flowerbuds. 1 cup	120	25	3	Trace	—	—	—	5	25	.8	70	.11	.10	.7	66
Celery, raw:															
Stalk, large outer, 8 by about 1½ inches, at root end. 1 stalk	40	5	Trace	Trace	—	—	—	2	16	.1	100	.01	.01	.1	4
Pieces, diced 1 cup	100	15	1	Trace	—	—	—	4	39	.3	240	.03	.03	.3	9
Collards, cooked 1 cup	190	55	5	1	—	—	—	9	239	1.1	10,260	.27	.37	2.4	87
Corn, sweet:															

Cooked, ear 5 by 1¾ inches.[4]															
1 ear	140	70	3	1	—	—	—	16	2	.5	310[5]	.09	.08	1.0	7
Canned, solids and liquid.															
1 cup	256	170	5	2	—	—	—	40	10	1.0	690[5]	.07	.12	2.3	13
Cowpeas, cooked, immature seeds.															
1 cup	160	175	13	1	—	—	—	29	38	3.4	560	.49	.18	2.3	28
Cucumbers, 10-ounce; 7½ by about 2 inches:															
Raw, pared															
1 cucumber	207	30	1	Trace	—	—	—	7	35	.6	Trace	.07	.09	.4	23
Raw, pared, center slice ⅛-inch thick.															
6 slices	50	5	Trace	Trace	—	—	—	2	8	.2	Trace	.02	.02	.1	6
Dandelion greens, cooked.															
1 cup	180	60	4	1	—	—	—	12	252	3.2	21,060	.24	.29	—	32
Endive, curly (including escarole).															
2 ounces	57	10	1	Trace	—	—	—	2	46	1.0	1,870	.04	.08	.3	6
Kale, leaves including stems, cooked.															
1 cup	110	30	4	1	—	—	—	4	147	1.3	8,140	—	—	—	68
Lettuce, raw:															
Butterhead, as Boston types; head, 4-inch diameter.															
1 head	220	30	3	Trace	—	—	—	6	77	4.4	2,130	.14	.13	.6	18
Crisphead, as iceberg; head, 4¾-inch diameter.															
1 head	454	60	4	Trace	—	—	—	13	91	2.3	1,500	.29	.27	1.3	29
Looseleaf, or bunching varieties, leaves.															
2 large	50	10	1	Trace	—	—	—	2	34	.7	950	.03	.04	.2	9
Mushrooms, canned, solids and liquid.															
1 cup	244	40	5	Trace	—	—	—	6	15	1.2	Trace	.04	.60	4.8	4
Mustard greens, cooked															
1 cup	140	35	3	1	—	—	—	6	193	2.5	8,120	.11	.19	.9	68
Okra, cooked, pod by ⅝-inch.															
8 pods	85	25	2	Trace	—	—	—	5	78	.4	420	.11	.15	.8	17

[4]Measure and weight apply to entire vegetable or fruit including parts not usually eaten.

[5]Based on yellow varieties; white varieties contain only a trace of cryptoxanthin and carotenes, the pigments in corn that have biological activity.

Food, Approximate Measure, and Weight (in Grams)		Food Energy	Protein	Fat (Total Lipid)	Fatty acids: Saturated (Total)	Fatty acids: Unsaturated Oleic	Fatty acids: Unsaturated Linoleic	Carbohydrate	Calcium	Iron	Vitamin A Value	Thiamine	Riboflavin	Niacin	Ascorbic Acid
	Grams	(Calories)	(Gm.)	(Gm.)	(Gm.)	(Gm.)	(Gm.)	(Gm.)	(Mo.)	(Mo.)	(I.U.)	(Mo.)	(Mo.)	(Mo.)	(Mo.)
VEGETABLES AND VEGETABLE PRODUCTS—Continued															
Onions:															
Mature:															
Raw, onion 2½-inch diameter. 1 onion	110	40	2	Trace	—	—	—	10	30	0.6	40	0.04	0.04	0.2	11
Cooked 1 cup	210	60	3	Trace	—	—	—	14	50	.8	80	.06	.06	.4	14
Young green, small, without tops. 6 onions	50	20	1	Trace	—	—	—	5	20	.3	Trace	.02	.02	.2	12
Parsley, raw, chopped 1 tablespoon	3.5	1	Trace	Trace	—	—	—	Trace	7	.2	300	Trace	.01	Trace	6
Parsnips, cooked 1 cup	155	100	2	1	—	—	—	23	70	.9	50	.11	.13	.2	16
Peas, green:															
Cooked 1 cup	160	115	9	1	—	—	—	19	37	2.9	860	.44	.17	3.7	33
Canned, solids and liquid. 1 cup	249	165	9	1	—	—	—	31	50	4.2	1,120	.23	.13	2.2	22
Canned, strained (baby food). 1 ounce	28	15	1	Trace	—	—	—	3	3	.4	140	.02	.02	.4	3
Peppers, hot, red, without seeds, dried (ground chili powder, added seasonings). 1 tablespoon	15	50	2	2	—	—	—	8	40	2.3	9,750	.03	.17	1.3	2
Peppers, sweet:															
Raw, medium, about 6 per pound:															

Green pod without stem and seeds.															
1 pod	62	15	1	Trace	—	—	—	3	6	.4	260	.05	.05	.3	79
Red pod without stem and seeds.															
1 pod	60	20	1	Trace	—	—	—	4	8	.4	2,670	.05	.05	.3	122
Canned, pimientos, medium.															
1 pod	38	10	Trace	Trace	—	—	—	2	3	.6	870	.01	.02	.1	36
Potatoes, medium (about 3 per pound raw):															
Baked, peeled after baking.															
1 potato	99	90	3	Trace	—	—	—	21	9	.7	Trace	.10	.04	1.7	20
Boiled:															
Peeled after boiling															
1 potato	136	105	3	Trace	—	—	—	23	10	.8	Trace	.13	.05	2.0	22
Peeled before boiling															
1 potato	122	80	2	Trace	—	—	—	18	7	.6	Trace	.11	.04	1.4	20
French-fried, piece 2 by ½ by ½-inch:															
Cooked in deep fat															
10 pieces	57	155	2	7	2	2	4	20	9	.7	Trace	.07	.04	1.8	12
Frozen, heated															
10 pieces	57	125	2	5	1	1	2	19	5	1.0	Trace	.08	.01	1.5	12
Mashed:															
Milk added															
1 cup	195	125	4	1	—	—	—	25	47	.8	50	.16	.10	2.0	19
Milk and butter added.															
1 cup	195	185	4	8	4	3	Trace	24	47	.8	330	.16	.10	1.9	18
Potato chips, medium, 2-inch diameter.															
10 chips	20	115	1	8	2	2	4	10	8	.4	Trace	.04	.01	1.0	3
Pumpkin, canned															
1 cup	228	75	2	1	—	—	—	18	57	.9	14,590	.07	.12	1.3	12
Radishes, raw, small, without tops.															
4 radishes	40	5	Trace	Trace	—	—	—	1	12	.4	Trace	.01	.01	.1	10
Sauerkraut, canned, solids and liquid.															
1 cup	235	45	2	Trace	—	—	—	9	85	1.2	120	.07	.09	.4	33
Spinach:															
Cooked															
1 cup	180	40	5	1	—	—	—	6	167	4.0	14,580	.13	.25	1.0	50
Canned, drained solids															
1 cup	180	45	5	1	—	—	—	6	212	4.7	14,400	.03	.21	.6	24

Food, Approximate Measure, and Weight (in Grams)		Food Energy	Protein	Fat (Total Lipid)	Fatty acids: Saturated (Total)	Fatty acids: Unsaturated: Oleic	Fatty acids: Unsaturated: Linoleic	Carbohydrate	Calcium	Iron	Vitamin A Value	Thiamine	Riboflavin	Niacin	Ascorbic Acid
	Grams	(Calories)	(Gm.)	(Gm.)	(Gm.)	(Gm.)	(Gm.)	(Gm.)	(Mg.)	(Mg.)	(I.U.)	(Mg.)	(Mg.)	(Mg.)	(Mg.)
Canned, strained or chopped (baby food). 1 ounce	28	10	1	Trace	—	—	—	2	18	.2	1,420	.01	.04	.1	2
Sprouts, raw:															
Mung bean 1 cup	90	30	3	Trace	—	—	—	6	17	1.2	20	.12	.12	.7	17
Soybean 1 cup	107	40	6	2	—	—	—	4	46	.7	90	.17	.16	.8	4
Squash:															
Cooked:															
Summer, diced 1 cup	210	30	2	Trace	—	—	—	7	52	.8	820	.10	.16	1.6	21
Winter, baked, mashed. 1 cup	205	130	4	1	—	—	—	32	57	1.6	8,610	.10	.27	1.4	27
Canned, winter, strained and chopped (baby food). 1 ounce	28	10	Trace	Trace	—	—	—	2	7	.1	510	.01	.01	.1	1
Sweet potatoes:															
Cooked, medium, 5 by 2 inches, weight raw about 6 ounces:															
Baked, peeled after baking. 1 sweet potato	110	155	2	1	—	—	—	36	44	1.0	8,910	0.10	0.07	0.7	24
Boiled, peeled after boiling. 1 sweet potato	147	170	2	1	—	—	—	39	47	1.0	11,610	.13	.09	.9	25
Candied, 3½ by 2¼ inches. 1 sweet potato	175	295	2	6	2	3	1	60	65	1.6	11,030	.10	.08	.8	17
Canned, vacuum or solid pack. 1 cup	218	235	4	Trace	—	—	—	54	54	1.7	17,000	.10	.10	1.4	30

Tomatoes:															
Raw, medium, 2 by 2½ inches, about 3 per pound.															
1 tomato	150	35	2	Trace	—	—	—	7	20	.8	1,350	.10	.06	1.0	34[6]
Canned															
1 cup	242	50	2	Trace	—	—	—	10	15	1.2	2,180	.13	.07	1.7	40
Tomato juice, canned															
1 cup	242	45	2	Trace	—	—	—	10	17	2.2	1,940	.13	.07	1.8	39
Tomato catsup															
1 tablespoon	17	15	Trace	Trace	—	—	—	4	4	.1	240	.02	.01	.3	3
Turnips, cooked, diced															
1 cup	155	35	1	Trace	—	—	—	8	54	.6	Trace	.06	.08	.5	33
Turnip greens:															
Cooked:															
In small amount of water, short time.															
1 cup	145	30	3	Trace	—	—	—	5	267	1.6	9,140	.21	.36	.8	100
In large amount of water, long time.															
1 cup	145	25	3	Trace	—	—	—	5	252	1.4	8,260	.14	.33	.8	68
Canned, solids and liquid															
1 cup	232	40	3	1	—	—	—	7	232	3.7	10,900	.04	.21	1.4	44
FRUITS AND FRUIT PRODUCTS															
Apples, raw, medium, 2½-inch diameter, about 3 per pound.[4]															
1 apple	150	70	Trace	Trace	—	—	—	18	8	.4	50	.04	.02	.1	3
Apple brown betty															
1 cup	230	345	4	8	4	3	Trace	68	41	1.4	230	.13	.10	.9	3
Apple juice, bottled or canned.															
1 cup	249	120	Trace	Trace	—	—	—	30	15	1.5	—	.01	.04	.2	2
Applesauce, canned:															
Sweetened															
1 cup	254	230	1	Trace	—	—	—	60	10	1.3	100	.05	.03	.1	3
Unsweetened or artificially sweetened.															
1 cup	239	100	Trace	Trace	—	—	—	26	10	1.2	100	.04	.02	.1	2

[6]Year-round average. Samples marketed from November through May average around 15 milligrams per 150-gram tomato; from June through October, around 39 milligrams.

Food, Approximate Measure, and Weight (in Grams)		Food Energy	Protein	Fat (Total Lipid)	Fatty acids: Saturated (Total)	Fatty acids: Unsaturated Oleic	Fatty acids: Unsaturated Linoleic	Carbohydrate	Calcium	Iron	Vitamin A Value	Thiamine	Riboflavin	Niacin	Ascorbic Acid
	Grams	(Calories)	(Gm.)	(Gm.)	(Gm.)	(Gm.)	(Gm.)	(Gm.)	(Mg.)	(Mg.)	(I.U.)	(Mg.)	(Mg.)	(Mg.)	(Mg.)
Applesauce and apricots, canned, strained or junior (baby food). 1 ounce	28	25	Trace	Trace	—	—	—	6	1	.1	170	Trace	Trace	Trace	1
Apricots:															
Raw, about 12 per pound.[4] 3 apricots	114	55	1	Trace	—	—	—	14	18	.5	2,890	.03	.04	.7	10
Canned in heavy sirup															
Halves and sirup 1 cup	259	220	2	Trace	—	—	—	57	28	.8	4,510	.05	.06	.9	10
Halves (medium) and sirup. 4 halves; 2 tablespoons sirup	122	105	1	Trace	—	—	—	27	13	.4	2,120	.02	.03	.4	5
Dried:															
Uncooked, 40 halves, small. 1 cup	150	390	8	1	—	—	—	100	100	8.2	16,350	.02	.23	4.9	19
Cooked, unsweetened, fruit and liquid. 1 cup	285	240	5	1	—	—	—	62	63	5.1	8,550	.01	.13	2.8	8
Apricot nectar, canned 1 cup	250	140	1	Trace	—	—	—	36	22	.5	2,380	.02	.02	.5	7
Avocados, raw:															
California varieties, mainly Fuerte:															
10-ounce avocado, about 3⅓ by 4¼ inches, peeled, pitted. ½ avocado	108	185	2	18	4	8	2	6	11	.6	310	.12	.21	1.7	15
½-inch cubes 1 cup	152	260	3	26	5	12	3	9	15	.9	440	.16	.30	2.4	21

[4]Measure and weight apply to entire vegetable or fruit including parts not usually eaten.

Food															
Florida varieties:															
13-ounce avocado, about 4 by 3 inches, peeled, pitted.															
½ avocado	123	160	2	14	3	6	2	11	12	.7	360	.13	.24	2.0	17
½-inch cubes															
1 cup	152	195	2	17	3	8	2	13	15	.9	440	.16	.30	2.4	21
Bananas, raw, 6 by 1½ inches, about 3 per pound.[4]															
1 banana	150	85	1	Trace	—	—	—	23	8	.7	190	.05	.06	.7	10
Blackberries, raw															
1 cup	144	85	2	1	—	—	—	19	46	1.3	290	.05	.06	.5	30
Blueberries, raw															
1 cup	140	85	1	1	—	—	—	21	21	1.4	140	.04	.08	.6	20
Cantaloupes, raw; medium, 5-inch diameter, about 1⅔ pounds.[4]															
½ melon	385	60	1	Trace	—	—	—	14	27	.8	6,540[7]	.08	.06	1.2	63
Cherries:															
Raw, sweet, with stems[4]															
1 cup	130	80	2	Trace	—	—	—	20	26	0.5	130	0.06	0.07	0.5	12
Canned, red, sour, pitted, heavy sirup.															
1 cup	260	230	2	1	—	—	—	59	36	.8	1,680	.07	.06	.4	13
Cranberry juice cocktail, canned.															
1 cup	250	160	Trace	Trace	—	—	—	41	12	.8	Trace	.02	.02	.1	(8)
Cranberry sauce, sweetened, canned, strained.															
1 cup	277	405	Trace	1	—	—	—	104	17	.6	40	.03	.03	.1	5
Dates, domestic, natural and dry, pitted, cut.															
1 cup	178	490	4	1	—	—	—	130	105	5.3	90	.16	.17	3.9	0
Figs:															
Raw, small, 1½-inch diameter, about 12 per pound.															
3 figs	114	90	1	Trace	—	—	—	23	40	.7	90	.07	.06	.5	2

[7]Value based on varieties with orange-colored flesh, for green-fleshed varieties value is about 540 I.U. per ½ melon.
[8]About 5 milligrams per 8 fluid ounces from cranberries. Ascorbic acid is usually added to approximately 100 milligrams per 8 fluid ounces.

Food, Approximate Measure, and Weight (in Grams)	Grams	Food Energy	Protein	Fat (Total Lipid)	Fatty acids: Saturated (Total)	Fatty acids: Unsaturated Oleic	Fatty acids: Unsaturated Linoleic	Carbohydrate	Calcium	Iron	Vitamin A Value	Thiamine	Riboflavin	Niacin	Ascorbic Acid
		(Calories)	(Gm.)	(Gm.)	(Gm.)	(Gm.)	(Gm.)	(Gm.)	(Mo.)	(Mo.)	(I.U.)	(Mo.)	(Mo.)	(Mo.)	(Mo.)
Dried, large, 2 by 1 inch 1 fig	21	60	1	Trace	—	—	—	15	26	.6	20	.02	.02	.1	0
Fruit cocktail, canned in heavy sirup, solids and liquid. 1 cup	256	195	1	1	—	—	—	50	23	1.0	360	.04	.03	1.1	5
Grapefruit:															
Raw, medium, 4¼-inch diameter, size 64:															
White[4] ½ grapefruit	285	55	1	Trace	—	—	—	14	22	.6	10	.05	.02	.3	52
Pink or red[4] ½ grapefruit	285	60	1	Trace	—	—	—	15	23	.6	640	.05	.02	.3	52
Raw sections, white 1 cup	194	75	1	Trace	—	—	—	20	31	.8	20	.07	.03	.3	72
Canned, white:															
Sirup pack, solids and liquid. 1 cup	249	175	1	Trace	—	—	—	44	32	.7	20	.07	.04	.5	75
Water pack, solids and liquid. 1 cup	240	70	1	Trace	—	—	—	18	31	.7	20	.07	.04	.5	72
Grapefruit juice:															
Fresh 1 cup	246	95	1	Trace	—	—	—	23	22	.5	(9)	.09	.04	.4	92
Canned, white:															
Unsweetened 1 cup	247	100	1	Trace	—	—	—	24	20	1.0	20	.07	.04	.4	84
Sweetened 1 cup	250	130	1	Trace	—	—	—	32	20	1.0	20	.07	.04	.4	78

[9]For white-fleshed varieties value is about 20 I.U. per cup; for red-fleshed varieties, 1,080 I.U. per cup.

Frozen, concentrate, unsweetened:														
Undiluted, can, 6 fluid ounces.														
1 can	207	300	4	1	—	—	—	72	70	.8	60	.29	.12	1.4	286
Diluted with 3 parts water, by volume.														
1 cup	247	100	1	Trace	—	—	—	24	25	.2	20	.10	.04	.5	96
Frozen, concentrate, sweetened:														
Undiluted, can, 6 fluid ounces.														
1 can	211	350	3	1	—	—	—	85	59	.6	50	.24	.11	1.2	245
Diluted with 3 parts water, by volume.														
1 cup	249	115	1	Trace	—	—	—	28	20	.2	20	.08	.03	.4	82
Dehydrated:														
Crystals, can, net weight 4 ounces.														
1 can	114	430	5	1	—	—	—	103	99	1.1	90	.41	.18	2.0	399
Prepared with water (1 pound yields about 1 gallon).														
1 cup	247	100	1	Trace	—	—	—	24	22	.2	20	.10	.05	.5	92
Grapes, raw:														
American type (slip skin), such as concord, Delaware, Niagara, Catawba, and Scuppernong[4]														
1 cup	153	65	1	1	—	—	—	15	15	.4	100	.05	.03	.2	3
European type (adherent skin), such as Malaga, Muscat, Thompson Seedless, Emperor, and Flame Tokay.[4]														
1 cup	160	95	1	Trace	—	—	—	25	17	.6	140	.07	.04	.4	6
Grape juice, bottled or canned.														
1 cup	254	165	1	Trace	—	—	—	42	28	.8	—	.10	.05	.6	Trace
Lemons, raw medium, 2⅓-inch diameter, size 150.[4]														
1 lemon	106	20	1	Trace	—	—	—	6	18	.4	10	.03	.01	.1	38

[4]Measure and weight apply to entire vegetable or fruit including parts not usually eaten.

Food, Approximate Measure, and Weight (in Grams)		Food Energy	Protein	Fat (Total Lipid)	Fatty acids: Saturated (Total)	Fatty acids: Unsaturated: Oleic	Fatty acids: Unsaturated: Linoleic	Carbohydrate	Calcium	Iron	Vitamin A Value	Thiamine	Riboflavin	Niacin	Ascorbic Acid
	Grams	(Calories)	(Gm.)	(Gm.)	(Gm.)	(Gm.)	(Gm.)	(Gm.)	(Mo.)	(Mo.)	(I.U.)	(Mo.)	(Mo.)	(Mo.)	(Mo.)
Lemon juice:															
Fresh															
1 cup	246	60	1	Trace	—	—	—	20	17	.5	40	.08	.03	.2	113
1 tablespoon	15	5	Trace	Trace	—	—	—	1	1	Trace	Trace	Trace	Trace	Trace	7
Canned, unsweetened															
1 cup	245	55	1	Trace	—	—	—	19	17	.5	40	.07	.03	.2	102
Lemonade concentrate, frozen, sweetened:															
Undiluted, can, 6 fluid ounces.															
1 can	220	430	Trace	Trace	—	—	—	112	9	.4	40	.05	.06	.7	66
Diluted with 4½ parts water, by volume.															
1 cup	248	110	Trace	Trace	—	—	—	28	2	.1	10	.01	.01	.2	17
Lime juice:															
Fresh															
1 cup	246	65	1	Trace	—	—	—	22	22	0.5	30	0.05	0.03	0.3	80
Canned															
1 cup	246	65	1	Trace	—	—	—	22	22	.5	30	.05	.03	.3	52
Limeade concentrate, frozen, sweetened:															
Undiluted, can, 6 fluid ounces.															
1 can	218	410	Trace	Trace	—	—	—	108	11	.2	Trace	.02	.02	.2	26
Diluted with 4⅓ parts water, by volume.															
1 cup	248	105	Trace	Trace	—	—	—	27	2	Trace	Trace	Trace	Trace	Trace	6
Oranges, raw:															
California, Navel (winter), 2⅘-inch diameter, size 88.[4]															

1 orange	180	60	2	Trace	—	—	—	16	49	.5	240	.12	.05	.5	75
Florida, all varieties, 3-inch diameter.[4]														
1 orange	210	75	1	Trace	—	—	—	19	67	.3	310	.16	.06	.6	70
Orange juice:														
Fresh:														
California, Valencia (summer).														
1 cup	249	115	2	1	—	—	—	26	27	.7	500	.22	.06	.9	122
Florida varieties:														
Early and midseason.														
1 cup	247	100	1	Trace	—	—	—	23	25	.5	490	.22	.06	.9	127
Late season, Valencia.														
1 cup	248	110	1	Trace	—	—	—	26	25	.5	500	.22	.06	.9	92
Canned, unsweetened														
1 cup	249	120	2	Trace	—	—	—	28	25	1.0	500	.17	.05	.6	100
Frozen concentrate:														
Undiluted, can, 6 fluid ounces.														
1 can	210	330	5	Trace	—	—	—	80	69	.8	1,490	.63	.10	2.4	332
Diluted with 3 parts water, by volume.														
1 cup	248	110	2	Trace	—	—	—	27	22	.2	500	.21	.03	.8	112
Dehydrated:														
Crystals, can, net weight 4 ounces.														
1 can	113	430	6	2	—	—	—	100	95	1.9	1,900	.76	.24	3.3	406
Prepared with water, 1 pound yields about 1 gallon.														
1 cup	248	115	1	Trace	—	—	—	27	25	.5	500	.20	.06	.9	108
Orange and grapefruit juice:														
Frozen concentrate:														
Undiluted, can, 6 fluid ounces.														
1 can	209	325	4	1	—	—	—	78	61	.8	790	.47	.06	2.3	301
Diluted with 3 parts water, by volume.														
1 cup	248	110	1	Trace	—	—	—	26	20	.2	270	.16	.02	.8	102
Papayas, raw, ½-inch cubes.														
1 cup	182	70	1	Trace	—	—	—	18	36	.5	3,190	.07	.08	.5	102

[4]Measure and weight apply to entire vegetable or fruit including parts not usually eaten.

Food, Approximate Measure, and Weight (in Grams)		Food Energy	Protein	Fat (Total Lipid)	Fatty acids: Saturated (Total)	Fatty acids: Unsaturated: Oleic	Fatty acids: Unsaturated: Linoleic	Carbohydrate	Calcium	Iron	Vitamin A Value	Thiamine	Riboflavin	Niacin	Ascorbic Acid
	Grams	(Calories)	(Gm.)	(Gm.)	(Gm.)	(Gm.)	(Gm.)	(Gm.)	(Mo.)	(Mo.)	(I.U.)	(Mo.)	(Mo.)	(Mo.)	(Mo.)
Peaches:															
Raw:															
Whole, medium, 2-inch diameter about 4 per pound.[4]															
1 peach	114	35	1	Trace	—	—	—	10	9	.5	1,320[10]	.02	.05	1.0	7
Sliced															
1 cup	168	65	1	Trace	—	—	—	16	15	.8	2,230[10]	.03	.08	1.6	12
Canned, yellow-fleshed, solids and liquid:															
Sirup pack, heavy:															
Halves or slices															
1 cup	257	200	1	Trace	—	—	—	52	10	.8	1,100	.02	.06	1.4	7
Halves (medium) and sirup.															
2 halves and 2 tablespoons sirup	117	90	Trace	Trace	—	—	—	24	5	.4	500	.01	.03	.7	3
Water pack															
1 cup	245	75	1	Trace	—	—	—	20	10	.7	1,000	.02	.06	1.4	7
Strained or chopped (baby food).															
1 ounce	28	25	Trace	Trace	—	—	—	6	2	.1	140	Trace	.01	.2	1
Dried:															
Uncooked															
1 cup	160	420	5	1	—	—	—	109	77	9.6	6,240	.02	.31	8.5	28
Cooked, unsweetened, 10-12 halves and 6 tablespoons liquid.															

[10]Based on yellow-fleshed varieties; for white-fleshed varieties value is about 50 I.U. per 114-gram peach and 80 I.U. per cup of sliced peaches.

1 cup	270	220	3	1	___	___	___	58	41	5.1	3,290	.01	.15	4.2	6
Frozen:															
Carton, 12 ounces, not thawed.															
1 carton	340	300	1	Trace	___	___	___	77	14	1.7	2,210	.03	.14	2.4	135[11]
Can, 16 ounces, not thawed.															
1 can	454	400	2	Trace	___	___	___	103	18	2.3	2,950	.05	.18	3.2	181[11]
Peach nectar, canned															
1 cup	250	120	Trace	Trace	___	___	___	31	10	.5	1,080	.02	.05	1.0	1
Pears:															
Raw, 3 by 2½-inch diameter.[4]															
1 pear	182	100	1	1	___	___	___	25	13	.5	30	.04	.07	.2	7
Canned, solids and liquid:															
Sirup pack, heavy:															
Halves or slices															
1 cup	255	195	1	1	___	___	___	50	13	0.5	Trace	0.03	0.05	0.3	4
Halves (medium) and sirup.															
2 halves and 2 tablespoons sirup	117	90	Trace	Trace	___	___	___	23	6	.2	Trace	.01	.02	.2	2
Water pack															
1 cup	243	80	Trace	Trace	___	___	___	20	12	.5	Trace	.02	.05	.3	4
Strained or chopped (baby food).															
1 ounce	28	20	Trace	Trace	___	___	___	5	2	.1	10	Trace	.01	.1	1
Pear nectar, canned															
1 cup	250	130	1	Trace	___	___	___	33	8	.2	Trace	.01	.05	Trace	1
Persimmons, Japanese or kaki, raw, seedless, 2½-inch diameter.[4]															
1 persimmon	125	75	1	Trace	___	___	___	20	6	.4	2,740	.03	.02	.1	11
Pineapple:															
Raw, diced															
1 cup	140	75	1	Trace	___	___	___	19	24	.7	100	.12	.04	.3	24
Canned, heavy sirup pack, solids and liquid:															
Crushed															
1 cup	260	195	1	Trace	___	___	___	50	29	.8	120	.20	.06	.5	17

[11]Average weighed in accordance with commercial freezing practices. For products without added ascorbic acid, value is about 37 milligrams per 12-ounce carton and 50 milligrams per 16-ounce can; for those with added ascorbic acid, 139 milligrams per 12 ounces and 186 milligrams per 16 ounces.

[4]Measure and weight apply to entire vegetable or fruit including parts not usually eaten.

Food, Approximate Measure, and Weight (in Grams)		Food Energy	Protein	Fat (Total Lipid)	Fatty acids: Saturated (Total)	Fatty acids: Unsaturated: Oleic	Fatty acids: Unsaturated: Linoleic	Carbohydrate	Calcium	Iron	Vitamin A Value	Thiamine	Riboflavin	Niacin	Ascorbic Acid
	Grams	(Calories)	(Gm.)	(Gm.)	(Gm.)	(Gm.)	(Gm.)	(Gm.)	(Mg.)	(Mg.)	(I.U.)	(Mg.)	(Mg.)	(Mg.)	(Mg.)
Sliced, slices and juice. 2 small or 1 large and 2 tablespoons juice	122	90	Trace	Trace	----	----	----	24	13	.4	50	.09	.03	.2	8
Pineapple juice, canned 1 cup	249	135	1	Trace	----	----	----	34	37	.7	120	.12	.04	.5	22
Plums, all except prunes:															
Raw, 2-inch diameter, about 2 ounces.[4] 1 plum	60	25	Trace	Trace	----	----	----	7	7	.3	140	.02	.02	.3	3
Canned, sirup pack (Italian prunes): Plums (with pits) and juice.[4] 1 cup	256	205	1	Trace	----	----	----	53	22	2.2	2,970	.05	.05	.9	4
Plums (without pits) and juice. 3 plums and 2 tablespoons juice	122	100	Trace	Trace	----	----	----	26	11	1.1	1,470	.03	.02	.5	2
Prunes, dried, "softenized," medium:															
Uncooked[4] 4 prunes	32	70	1	Trace	----	----	----	18	14	1.1	440	.02	.04	.4	1
Cooked, unsweetened, 17-18 prunes and 1/3 cup liquid.[4] 1 cup	270	295	2	1	----	----	----	78	60	4.5	1,860	.08	.18	1.7	2
Prunes with tapioca, canned, strained or junior (baby food). 1 ounce	28	25	Trace	Trace	----	----	----	6	2	.3	110	.01	.02	.1	1
Prune juice, canned 1 cup	256	200	1	Trace	----	----	----	49	36	10.5	----	.02	.03	1.1	4
Raisins, dried 1 cup	160	460	4	Trace	----	----	----	124	99	5.6	30	.18	.13	.9	2

Food															
Raspberries, red:															
Raw															
1 cup	123	70	1	1	—	—	—	17	27	1.1	160	.04	.11	1.1	31
Frozen, 10-ounce carton, not thawed.															
1 carton	284	275	2	1	—	—	—	70	37	1.7	200	.06	.17	1.7	59
Rhubarb, cooked, sugar added.															
1 cup	272	385	1	Trace	—	—	—	98	212	1.6	220	.06	.15	.7	17
Strawberries:															
Raw, capped															
1 cup	149	55	1	1	—	—	—	13	31	1.5	90	.04	.10	1.0	88
Frozen, 10-ounce carton, not thawed.															
1 carton	284	310	1	1	—	—	—	79	40	2.0	90	.06	.17	1.5	150
Frozen, 16-ounce can, not thawed.															
1 can	454	495	2	1	—	—	—	126	64	3.2	150	.09	.27	2.4	240
Tangerines, raw, medium, 2½-inch diameter, about 4 per pound.[4]															
1 tangerine	114	40	1	Trace	—	—	—	10	34	.3	350	.05	.02	.1	26
Tangerine juice:															
Canned, unsweetened															
1 cup	248	105	1	Trace	—	—	—	25	45	.5	1,040	.14	.04	.3	56
Frozen concentrate:															
Undiluted, can, 6 fluid ounces.															
1 can	210	340	4	1	—	—	—	80	130	1.5	3,070	.43	.12	.9	202
Diluted, with 3 parts water, by volume.															
1 cup	248	115	1	Trace	—	—	—	27	45	.5	1,020	.14	.04	.3	67
Watermelon, raw, wedge, 4 by 8 inches (1/16 of 10 by 16-inch melon, about 2 pounds with rind).[4]															
1 wedge	925	115	2	1	—	—	—	27	30	2.1	2,510	.13	.13	.7	30
GRAIN PRODUCTS															
Barley, pearled, light, uncooked.															
1 cup	203	710	17	2	Trace	1	1	160	32	4.1	0	0.25	0.17	6.3	0
Biscuits, baking powder with enriched flour, 2½-inch diameter.															
1 biscuit	38	140	3	6	2	3	1	17	46	.6	Trace	.08	.08	.7	Trace

[4]Measure and weight apply to entire vegetable or fruit including parts not usually eaten.

Food, Approximate Measure, and Weight (in Grams)		Food Energy	Protein	Fat (Total Lipid)	Fatty acids: Saturated (Total)	Fatty acids: Unsaturated: Oleic	Fatty acids: Unsaturated: Linoleic	Carbohydrate	Calcium	Iron	Vitamin A Value	Thiamine	Riboflavin	Niacin	Ascorbic Acid
	Grams	(Calories)	(Gm.)	(Gm.)	(Gm.)	(Gm.)	(Gm.)	(Gm.)	(Mo.)	(Mo.)	(I.U.)	(Mo.)	(Mo.)	(Mo.)	(Mo.)
Bran flakes (40 percent bran) added thiamine.															
1 ounce	28	85	3	1	—	—	—	23	20	1.2	0	.11	.05	1.7	0
Breads:															
Boston brown bread, slice, 3 by ¾ inch.															
1 slice	48	100	3	1	—	—	—	22	43	.9	0	.05	.03	.6	0
Cracked-wheat bread:															
Loaf, 1-pound, 20 slices.															
1 loaf	454	1190	39	10	2	5	2	236	399	5.0	Trace	.53	.42	5.8	Trace
Slice															
1 slice	23	60	2	1	—	—	—	12	20	.3	Trace	.03	.02	.3	Trace
French or Vienna bread:															
Enriched, 1-pound loaf.															
1 loaf	454	1315	41	14	3	8	2	251	195	10.0	Trace	1.26	.98	11.3	Trace
Unenriched, 1-pound loaf.															
1 loaf	454	1315	41	14	3	8	2	251	195	3.2	Trace	.39	.39	3.6	Trace
Italian bread:															
Enriched, 1-pound loaf.															
1 loaf	454	1250	41	4	Trace	1	2	256	77	10.0	0	1.31	.93	11.7	0
Unenriched, 1-pound loaf.															
1 loaf	454	1250	41	4	Trace	1	2	256	77	3.2	0	.39	.27	3.6	0
Raisin bread:															
Loaf, 1-pound, 20 slices.															
1 loaf	454	1190	30	13	3	8	2	243	322	5.9	Trace	.24	.42	3.0	Trace
Slice															
1 slice	23	60	2	1	—	—	—	12	16	.3	Trace	.01	.02	.2	Trace
Rye bread:															

American, light (1/3 rye, 2/3 wheat):															
Loaf, 1-pound, 20 slices.															
1 loaf	454	1100	41	5	—	—	—	236	340	7.3	0	.81	.33	6.4	0
Slice															
1 slice	23	55	2	Trace	—	—	—	12	17	.4	0	.04	.02	.3	0
Pumpernickel, loaf, 1 pound.															
1 loaf	454	1115	41	5	—	—	—	241	381	10.9	0	1.05	.63	5.4	0
White bread, enriched:															
1 to 2 percent nonfat dry milk:															
Loaf, 1-pound, 20 slices.															
1 loaf	454	1225	39	15	3	8	2	229	318	10.9	Trace	1.13	.77	10.4	Trace
Slice															
1 slice	23	60	2	1	Trace	Trace	Trace	12	16	.6	Trace	.06	.04	.5	Trace
3 to 4 percent nonfat dry milk:[12]															
Loaf, 1-pound															
1 loaf	454	1225	39	15	3	8	2	229	381	11.3	Trace	1.13	.95	10.8	Trace
Slice, 20 per loaf															
1 slice	23	60	2	1	Trace	Trace	Trace	12	19	.6	Trace	.06	.05	.6	Trace
Slice, toasted															
1 slice	20	60	2	1	Trace	Trace	Trace	12	19	.6	Trace	.05	.05	.6	Trace
Slice, 26 per loaf															
1 slice	17	45	1	1	Trace	Trace	Trace	9	14	.4	Trace	.04	.04	.4	Trace
5 to 6 percent nonfat dry milk:															
Loaf, 1-pound, 20 slices.															
1 loaf	454	1245	41	17	4	10	2	228	435	11.3	Trace	1.22	.91	11.0	Trace
Slice															
1 slice	23	65	2	1	Trace	Trace	Trace	12	22	.6	Trace	.06	.05	.6	Trace
White bread, unenriched:															
1 to 2 percent nonfat dry milk:															
Loaf, 1-pound, 20 slices.															
1 loaf	454	1225	39	15	3	8	2	229	318	3.2	Trace	.40	.36	5.6	Trace
Slice															
1 slice	23	60	2	1	Trace	Trace	Trace	12	16	.2	Trace	.02	.02	.3	Trace
3 to 4 percent nonfat dry milk:[12]															
Loaf, 1-pound															

[12]When the amount of nonfat dry milk in commercial white bread is unknown, values for bread with 3 to 4 percent nonfat dry milk are suggested.

Food, Approximate Measure, and Weight (in Grams)		Food Energy	Protein	Fat (Total Lipid)	Fatty acids: Saturated (Total)	Fatty acids: Unsaturated: Oleic	Fatty acids: Unsaturated: Linoleic	Carbohydrate	Calcium	Iron	Vitamin A Value	Thiamine	Riboflavin	Niacin	Ascorbic Acid
	Grams	(Calories)	(Gm.)	(Gm.)	(Gm.)	(Gm.)	(Gm.)	(Gm.)	(Mg.)	(Mg.)	(I.U.)	(Mg.)	(Mg.)	(Mg.)	(Mg.)
1 loaf	454	1225	39	15	3	8	----	229	381	3.2	Trace	.31	.39	5.0	Trace
Slice, 20 per loaf															
1 slice	23	60	2	1	Trace	Trace	Trace	12	19	.2	Trace	.02	.02	.3	Trace
Slice, toasted															
1 slice	20	60	2	1	Trace	Trace	Trace	12	19	.2	Trace	.01	.02	.3	Trace
Slice, 26 per loaf															
1 slice	17	45	1	1	Trace	Trace	Trace	9	14	.1	Trace	.01	.01	.2	Trace
5 to 6 percent nonfat dry milk:															
Loaf, 1 pound, 20 slices.															
1 loaf	454	1245	41	17	4	10	2	228	435	3.2	Trace	.32	.59	4.1	Trace
Slice															
1 slice	23	65	2	1	Trace	Trace	Trace	12	22	.2	Trace	.02	.03	.2	Trace
Whole-wheat bread, made with 2 percent nonfat dry milk:															
Loaf, 1-pound, 20 slices															
1 loaf	454	1105	48	14	3	6	3	216	449	10.4	Trace	1.17	.56	12.9	Trace
Slice															
1 slice	23	55	2	1	Trace	Trace	Trace	11	23	.5	Trace	.06	.03	.7	Trace
Slice, toasted															
1 slice	19	55	2	1	Trace	Trace	Trace	11	22	.5	Trace	.05	.03	.6	Trace
Breadcrumbs, dry, grated															
1 cup	88	345	11	4	1	2	1	65	107	3.2	Trace	.19	.26	3.1	Trace
Cakes:[18]															
Angelfood cake; sector, 2-inch (1/12 of 8-inch-diameter cake).															
1 sector	40	110	3	Trace	----	----	----	24	4	.1	0	Trace	.06	.1	0

[18]Unenriched cake flour and vegetable cooking fat used unless otherwise specified.

Food														
Cakes[13]—Continued														
Chocolate cake, chocolate icing; sector, 2-inch (1/16 of 10-inch-diameter layer cake).														
1 sector	120	445	5	20	8	10	1	67	84	1.2	[14]190	0.03	0.12	0.3	Trace
Fruitcake, dark (made with enriched (flour); piece, 2 by 2 by ½ inch.														
1 piece	30	115	1	5	1	3	1	18	22	.8	[14]40	.04	.04	.2	Trace
Gingerbread (made with enriched flour); piece, 2 by 2 by 2 inches.														
1 piece	55	175	2	6	1	4	Trace	29	37	1.3	50	.06	.06	.5	0
Plain cake and cupcakes, without icing:														
Piece, 3 by 2 by 1½ inches.														
1 piece	55	200	2	8	2	5	1	31	35	.2	[14]90	.01	.05	.1	Trace
Cupcake, 2¾-inch diameter.														
1 cupcake	40	145	2	6	1	3	Trace	22	26	.2	[14]70	.01	.03	.1	Trace
Plain cake and cupcakes, with chocolate icing:														
Sector, 2 inch (1/16 of 10-inch-layer cake).														
1 sector	100	370	4	14	5	7	1	59	63	.6	[14]180	.02	.09	.2	Trace
Cupcake, 2¾-inch diameter.														
1 cupcake	50	185	2	7	2	4	Trace	30	32	.3	[14]90	.01	.04	.1	Trace
Poundcake, oldfashioned (equal weights flour, sugar, fat, eggs); slice, 2¾ by 3 by ⅝ inch.														
1 slice	30	140	2	9	2	5	1	14	6	.2	[14]80	.01	.03	.1	0
Sponge cake; sector, 2-inch (1/12 of 8-inch-diameter cake).														
1 sector	40	120	3	2	1	1	Trace	22	12	.5	180	.02	.06	.1	Trace

[13]Unenriched cake flour and vegetable cooking fat used unless otherwise specified.

[14]If the fat used in the recipe is butter or fortified margarine, the vitamin A value for chocolate cake with chocolate icing will be 490 I.U. per 2-inch, 100 I.U. for fruitcake, for plain cake without icing, 300 I.U. per piece, 220 I.U. per cupcake, for plain cake with icing, 440 I.U. per 2-inch sector, 220 I.U. per cupcake, and 300 I.U. for poundcake.

Food, Approximate Measure, and Weight (in Grams)		Food Energy	Protein	Fat (Total Lipid)	Fatty acids: Saturated (Total)	Fatty acids: Unsaturated: Oleic	Fatty acids: Unsaturated: Linoleic	Carbohydrate	Calcium	Iron	Vitamin A Value	Thiamine	Riboflavin	Niacin	Ascorbic Acid
	Grams	(Calories)	(Gm.)	(Gm.)	(Gm.)	(Gm.)	(Gm.)	(Gm.)	(Mg.)	(Mg.)	(I.U.)	(Mg.)	(Mg.)	(Mg.)	(Mg.)
Cookies:															
Plain and assorted. 3-inch diameter.															
1 cooky	25	120	1	5	—	—	—	18	9	.2	20	.01	.01	.1	Trace
Fig bars, small															
1 fig bar	16	55	1	1	—	—	—	12	12	.2	20	.01	.01	.1	Trace
Corn, rice and wheat flakes, mixed, added nutrients.															
1 ounce	28	110	2	Trace	—	—	—	24	11	.5	0	.11	—	.9	0
Corn flakes, added nutrients:															
Plain															
1 ounce	28	110	2	Trace	—	—	—	24	5	.4	0	.12	.02	.6	0
Sugar-covered															
1 ounce	28	110	1	Trace	—	—	—	26	3	.3	0	.12	.01	.5	0
Corn grits, degermed, cooked:															
Enriched															
1 cup	242	120	3	Trace	—	—	—	27	2	.7[15]	150[16]	.10[15]	.07[15]	1.0[15]	0
Unenriched															
1 cup	242	120	3	Trace	—										
Cornmeal, white or yellow, dry:						—	—	27	2	.2	150[16]	.05	.02	.5	0
Whole ground, unbolted															
1 cup	118	420	11	5	1	2	2	87	24	2.8	600[16]	.45	.13	2.4	0
Degermed, enriched															
1 cup	145	525	11	2	Trace	1	1	114	9	4.2[15]	640[16]	.64[15]	.38[15]	5.1[15]	0

[15]Iron, thiamine, riboflavin, and niacin are based on the minimum levels of enrichment specified in standards of identity promulgated under the Federal Food, Drug, and Cosmetic Act.

Food															
Corn muffins, made with enriched degermed cornmeal and enriched flour; muffin, 2¾-inch diameter. 1 muffin	48	150	3	5	2	2	Trace	23	50	.8	80[17]	.09	.11	.8	Trace
Corn, puffed, presweetened, added nutrients. 1 ounce	28	110	1	Trace	——	——	——	26	3	.5	0	.12	.05	.6	0
Corn, shredded, added nutrients. 1 ounce	28	110	2	Trace	——	——	——	25	1	.7	0	.12	.05	.6	0
Crackers:															
Graham, plain 4 small or 2 medium	14	55	1	1	——	——	——	10	6	.2	0	.01	.03	.2	0
Saltines, 2 inches square. 2 crackers	8	35	1	1	——	——	——	6	2	.1	0	Trace	Trace	.1	0
Soda:															
Cracker, 2½ inches square. 2 crackers	11	50	1	1	Trace	1	Trace	8	2	.2	0	Trace	Trace	.1	0
Oyster crackers 10 crackers	10	45	1	1	Trace	1	Trace	7	2	.2	0	Trace	Trace	.1	0
Cracker meal 1 tablespoon	10	45	1	1	Trace	1	Trace	7	2	.1	0	.01	Trace	.1	0
Doughnuts, cake type 1 doughnut	32	125	1	6	1	4	Trace	16	13	.4[18]	30	.05[18]	.05[18]	.4[18]	Trace
Farina, regular, enriched, cooked. 1 cup	238	100	3	Trace	——	——	——	21	10	.7[15]	0	.11[15]	.07[15]	1.0[15]	0

[16]Vitamin A value based on yellow product. White product contains only a trace.

[17]Based on recipe using white cornmeal; if yellow cornmeal is used, the vitamin A value is 140 I.U. per muffin.

[18]Based on product made with enriched flour. With unenriched flour, approximate values per doughnut are: iron, 0.2 milligram; thiamine, 0.01 milligram; riboflavin, 0.03 milligram; niacin, 0.2 milligram.

Food, Approximate Measure, and Weight (in Grams)		Food Energy	Protein	Fat (Total Lipid)	Fatty acids: Saturated (Total)	Fatty acids: Unsaturated		Carbohydrate	Calcium	Iron	Vitamin A Value	Thiamine	Riboflavin	Niacin	Ascorbic Acid
						Oleic	Linoleic								
	Grams	(Calories)	(Gm.)	(Gm.)	(Gm.)	(Gm.)	(Gm.)	(Gm.)	(Mg.)	(Mg.)	(I.U.)	(Mg.)	(Mg.)	(Mg.)	(Mg.)
Macaroni, cooked:															
Enriched:															
Cooked, firm stage (8 to 10 minutes; undergoes additional cooking in a food mixture). 1 cup	130	190	6	1	———	———	———	39	14	1.4[15]	0	0.23[15]	0.14[15]	1.9[15]	0
Cooked until tender 1 cup	140	155	5	1	———	———	———	32	11	1.8[15]	0	.19[15]	.11[15]	1.5[15]	0
Unenriched:															
Cooked, firm stage (8 to 10 minutes; undergoes additional cooking in a food mixture). 1 cup	130	190	6	1	———	———	———	39	14	.6	0	.02	.02	.5	0
Cooked until tender 1 cup	140	155	5	1	———	———	———	32	11	.6	0	.02	.02	.4	0
Macaroni (enriched) and cheese, baked. 1 cup	220	470	18	24	11	10	1	44	398	2.0	950	.22	.44	2.0	Trace
Muffins, with enriched white flour; muffin, 2¾-inch diameter. 1 muffin	48	140	4	5	1	3	Trace	20	50	.8	50	.08	.11	.7	Trace
Noodles (egg noodles), cooked:															
Enriched 1 cup	160	200	7	2	1	1	Trace	37	16	1.4[15]	110	.23[15]	.14[15]	1.8[15]	0
Unenriched 1 cup	160	200	7	2	1	1	Trace	37	16	1.0	110	.04	.03	.7	0

[15]Iron, thiamine, riboflavin, and niacin are based on the minimum levels of enrichment specified in standards of identity promulgated under the Federal Food, Drug, and Cosmetic Act.

Food															
Oats (with or without corn) puffed, added nutrients.															
1 ounce	28	115	3	2	Trace	1	1	21	50	1.3	0	.28	.05	.5	0
Oatmeal or rolled oats, regular or quick-cooking, cooked.															
1 cup	236	130	5	2	Trace	1	1	23	21	1.4	0	1.9	.05	.3	0
Pancakes (griddlecakes), 4-inch diameter:															
Wheat, enriched flour (home recipe).															
1 cake	27	60	2	2	Trace	1	Trace	9	27	.4	30	.05	.06	.3	Trace
Buckwheat (buckwheat pancake mix, made with egg and milk).															
1 cake	27	55	2	2	1	1	Trace	6	59	.4	60	.03	.04	.2	Trace
Piecrust, plain, baked:															
Enriched flour:															
Lower crust, 9-inch shell.															
1 crust	135	675	8	45	10	20	3	59	19	2.3	0	.27	.19	2.4	0
Double crust, 9-inch pie.															
1 double crust	270	1350	16	90	21	58	7	118	38	4.6	0	.55	.39	4.9	0
Unenriched flour:															
Lower crust, 9-inch shell.															
1 crust	135	675	8	45	10	29	3	59	19	.7	0	.04	.04	.6	0
Double crust, 9-inch pie.															
1 double crust	270	1350	16	90	21	58	7	118	38	1.4	0	.08	.07	1.3	0
Pies (piecrust made with unenriched flour); sector, 4-inch, 1/7 of 9-inch-diameter pie:															
Apple															
1 sector	135	345	3	15	4	9	1	51	11	.4	40	.03	.02	.5	1
Cherry															
1 sector	135	355	4	15	4	10	1	52	19	.4	590	.03	.03	.6	1
Custard															
1 sector	130	280	8	14	5	8	1	30	125	.8	300	.07	.21	.4	0
Mince															
1 sector	135	365	3	16	4	10	1	56	38	1.4	Trace	.09	.05	.5	1
Pumpkin															
1 sector	130	275	5	15	5	7	1	32	66	.6	3,210	.04	.13	.6	Trace

Food, Approximate Measure, and Weight (in Grams)		Food Energy	Protein	Fat (Total Lipid)	Fatty acids: Saturated (Total)	Fatty acids: Unsaturated		Carbohydrate	Calcium	Iron	Vitamin A Value	Thiamine	Riboflavin	Niacin	Ascorbic Acid
						Oleic	Linoleic								
	Grams	(Calories)	(Gm.)	(Gm.)	(Gm.)	(Gm.)	(Gm.)	(Gm.)	(Mo.)	(Mo.)	(I.U.)	(Mo.)	(Mo.)	(Mo.)	(Mo.)
Pizza (cheese); 5½-inch sector; ⅛ of 14-inch-diameter pie.															
1 sector	75	185	7	6	2	3	Trace	27	107	.7	290	.04	.12	.7	4
Popcorn, popped, with added oil and salt.															
1 cup	14	65	1	3	2	Trace	Trace	8	1	.3	—	—	.01	.2	0
Pretzels, small stick															
5 sticks	5	20	Trace	Trace	—	—	—	4	1	0	0	Trace	Trace	Trace	0
Rice, white (fully milled or polished), enriched, cooked:															
Common commercial varieties, all types.															
1 cup	168	185	3	Trace	—	—	—	41	17	1.3[15]	0	1.9[15]	.11[15]	1.5[15]	0
Long grain, parboiled															
1 cup	176	185	4	Trace	—	—	—	41	33	1.5[19]	0	.19[19]	.01[19]	1.6[19]	0
Rice, puffed, added nutrients (without salt).															
1 cup	14	55	1	Trace	—	—	—	13	3	.3	0	.06	.01	.6	0
Rice flakes, added nutrients.															
1 cup	30	115	2	Trace	—	—	—	26	9	0.5	0	0.10	0.02	1.6	0
Rolls:															
Plain, pan; 12 per 16 ounces:															
Enriched															

[19]Iron, thiamine, and niacin are based on the minimum levels of enrichment specified in standards of identity promulgated under the Federal Food, Drug, and Cosmetic Act. Riboflavin is based on unenriched rice. When the minimum level of enrichment for riboflavin specified in the standards of identity becomes effective the value will be 0.12 milligram per cup of parboiled rice and of white rice.

1 roll	38	115	3	2	Trace	1	Trace	20	28	.7	Trace	.11	.07	.8	Trace
Unenriched															
1 roll	38	115	3	2	Trace	1	Trace	20	28	.3	Trace	.02	.03	.3	Trace
Hard, round; 12 per 22 ounces.															
1 roll	52	160	5	2	Trace	1	Trace	31	24	.4	Trace	.03	.05	.4	Trace
Sweet, pan; 12 per 18 ounces.															
1 roll	43	135	4	4	1	2	Trace	21	37	.3	30	.03	.06	.4	Trace
Rye wafers, whole-grain, 1⅞ by 3½ inches.															
2 wafers	13	45	2	Trace	—	—	—	10	7	.5	0	.04	.03	.2	0
Spaghetti:															
Cooked, tender stage (14 to 20 minutes):															
Enriched															
1 cup	140	155	5	1	—	—	—	32	11	1.4[19]	0	.19[19]	.02[19]	2.0[19]	0
Unenriched															
1 cup	140	155	5	1	—	—	—	32	11	.6	0	.02	.02	.4	0
Spaghetti with meat balls in tomato sauce (home recipe).															
1 cup	250	335	19	12	4	6	1	39	125	3.8	1,600	.26	.30	4.0	22
Spaghetti in tomato sauce with cheese (home recipe).															
1 cup	250	260	9	9	2	5	1	37	80	2.2	1,080	.24	.18	2.4	14
Waffles, with enriched flour, ½ by 4½ by 5½ inches.															
1 waffle	75	210	7	7	2	4	1	28	85	1.3	250	.13	.19	1.0	Trace
Wheat, puffed:															
With added nutrients (without salt).															
1 ounce	28	105	4	Trace	—	—	—	22	8	1.2	0	.15	.07	2.2	0
With added nutrients, with sugar and honey.															
1 ounce	28	105	2	1	—	—	—	25	7	.9	0	.14	.05	1.8	0
Wheat, rolled; cooked															
1 cup	236	175	5	1	—	—	—	40	19	1.7	0	.17	.06	2.1	0

[15]Iron, thiamine, riboflavin, and niacin are based on the minimum levels of enrichment specified in standards of identity promulgated under the Federal Food, Drug, and Cosmetic Act.

Food, Approximate Measure, and Weight (in Grams)		Food Energy	Protein	Fat (Total Lipid)	Fatty acids: Saturated (Total)	Unsaturated: Oleic	Unsaturated: Linoleic	Carbohydrate	Calcium	Iron	Vitamin A Value	Thiamine	Riboflavin	Niacin	Ascorbic Acid
	Grams	(Calories)	(Gm.)	(Gm.)	(Gm.)	(Gm.)	(Gm.)	(Gm.)	(Mg.)	(Mg.)	(I.U.)	(Mg.)	(Mg.)	(Mg.)	(Mg.)
Wheat, shredded, plain (long, round, or bitesize). 1 ounce	28	100	3	1	—	—	—	23	12	1.0	0	.06	.03	1.2	0
Wheat and malted barley flakes, with added nutrients. 1 ounce	28	110	2	Trace	—	—	—	24	14	.7	0	.13	.03	1.1	0
Wheat flakes, with added nutrients. 1 ounce	28	100	3	Trace	—	—	—	23	12	1.2	0	.18	.04	1.4	0
Wheat flours:															
Whole-wheat, from hard wheats, stirred. 1 cup	120	400	16	2	Trace	1	1	85	49	4.0	0	.66	.14	5.2	0
All-purpose or family flour:															
Enriched, sifted 1 cup	110	400	12	1	Trace	Trace	Trace	84	18	3.2[15]	0	.48[15]	.29[15]	3.8[15]	0
Unenriched, sifted 1 cup	110	400	12	1	Trace	Trace	Trace	84	18	.9	0	.07	.05	1.0	0
Self-rising, enriched 1 cup	110	385	10	1	Trace	Trace	Trace	82	292	3.2[15]	0	.49[15]	.29[15]	3.9[15]	0
Cake or pastry flour, sifted. 1 cup	100	365	8	1	Trace	Trace	Trace	79	17	.5	0	.03	.03	.7	0
Wheat germ, crude, commercially milled. 1 cup	68	245	18	7	1	2	4	32	49	6.4	0	1.36	.46	2.9	0

[15] Iron, thiamine, riboflavin, and niacin are based on the minimum levels of enrichment specified in standards of identity promulgated under the Federal Food, Drug, and Cosmetic Act.

FATS, OILS

Butter, 4 sticks per pound:															
Sticks, 2															
1 cup	277	1625	1	184	101	61	6	1	45	0	7,500[20]	—	—	—	0
Sticks, 1/8															
1 tablespoon	14	100	Trace	11	6	4	Trace	Trace	3	0	460[20]	—	—	—	0
Pat or square (64 per pound).															
1 pat	7	50	Trace	6	3	2	Trace	Trace	1	0	230[20]	—	—	—	0
Fats, cooking:															
Lard															
1 cup	200	1985	0	220	84	101	22	0	0	0	0	0	0	0	0
Lard															
1 tablespoon	14	125	0	14	5	6	1	0	0	0	0	0	0	0	0
Vegetable fats															
1 cup	200	1770	0	200	46	130	14	0	0	0	—	0	0	0	0
Vegetable fats															
1 tablespoon	12.5	110	0	12	3	8	1	0	0	0	—	0	0	0	0
Margarine, 4 sticks per pound:															
Sticks, 2															
1 cup	227	1635	1	184	37	105	33	1	45	0	7,500[21]	—	—	—	0
Stick, 1/8															
1 tablespoon	14	100	Trace	11	2	6	2	Trace	3	0	460[21]	—	—	—	0
Pat or square (64 per pound).															
1 pat	7	50	Trace	6	1	3	1	Trace	1	0	230[21]	—	—	—	0
Oils, salad or cooking:															
Corn															
1 tablespoon	14	125	0	14	1	4	7	0	0	0	—	0	0	0	0
Cottonseed															
1 tablespoon	14	125	0	14	4	3	7	0	0	0	—	0	0	0	0
Olive															
1 tablespoon	14	125	0	14	2	11	1	0	0	0	—	0	0	0	0
Soybean															
1 tablespoon	14	125	0	14	2	3	7	0	0	0	—	0	0	0	0
Salad dressings:															

[20]Year-round average.

[21]Based on the average vitamin A content of fortified margarine. Federal specifications for fortified margarine require a minimum of 15,000 I.U. of vitamin A per pound.

Food, Approximate Measure, and Weight (in Grams)		Food Energy	Protein	Fat (Total Lipid)	Fatty acids: Saturated (Total)	Fatty acids: Unsaturated: Oleic	Fatty acids: Unsaturated: Linoleic	Carbohydrate	Calcium	Iron	Vitamin A Value	Thiamine	Riboflavin	Niacin	Ascorbic Acid
	Grams	(Calories)	(Gm.)	(Gm.)	(Gm.)	(Gm.)	(Gm.)	(Gm.)	(Mg.)	(Mg.)	(I.U.)	(Mg.)	(Mg.)	(Mg.)	(Mg.)
Blue cheese 1 tablespoon	16	80	1	8	2	2	4	1	13	Trace	30	Trace	0.02	Trace	Trace
Commercial, mayonnaise type. 1 tablespoon	15	65	Trace	6	1	1	3	2	2	Trace	30	Trace	Trace	Trace	----
French 1 tablespoon	15	60	Trace	6	1	1	3	3	2	.1	----	----	----	----	----
Home cooked, boiled 1 tablespoon	17	30	1	2	1	1	Trace	3	15	.1	80	.01	.03	Trace	Trace
Mayonnaise 1 tablespoon	15	110	Trace	12	2	3	6	Trace	3	.1	40	Trace	.01	Trace	----
Thousand Island 1 tablespoon	15	75	Trace	8	1	2	4	2	2	.1	50	Trace	Trace	Trace	Trace
SUGARS, SWEETS															
Candy:															
Caramels 1 ounce	28	115	1	3	2	1	Trace	22	42	.4	Trace	.01	.05	Trace	Trace
Chocolate, milk, plain 1 ounce	28	150	2	9	5	3	Trace	16	65	.3	80	.02	.09	.1	Trace
Fudge, plain 1 ounce	28	115	1	3	2	1	Trace	21	22	.3	Trace	.01	.03	.1	Trace
Hard candy 1 ounce	28	110	0	Trace	----	----	----	28	6	.5	0	0	0	0	0
Marshmallows 1 ounce	28	90	1	Trace	----	----	----	23	5	.5	0	0	Trace	Trace	0
Chocolate sirup, thin type 1 tablespoon	20	50	Trace	Trace	Trace	Trace	Trace	13	3	.3	----	Trace	.01	.1	0

Food															
Honey, strained or extracted.															
1 tablespoon	21	65	Trace	0	—	—	—	17	1	.1	0	Trace	.01	.1	Trace
Jams and preserves															
1 tablespoon	20	55	Trace	Trace	—	—	—	14	4	.2	Trace	Trace	.01	Trace	Trace
Jellies															
1 tablespoon	20	55	Trace	Trace	—	—	—	14	4	.3	Trace	Trace	.01	Trace	1
Molasses, cane:															
Light (first extraction)															
1 tablespoon	20	50	—	—	—	—	—	13	33	.9	—	.01	.01	Trace	—
Blackstrap (third extraction).															
1 tablespoon	20	45	—	—	—	—	—	11	137	3.2	—	.02	.04	.4	—
Sirup, table blends (chiefly corn, light and dark).															
1 tablespoon	20	60	0	0	—	—	—	15	9	.8	0	0	0	0	0
Sugars (cane or beet):															
Granulated															
1 cup	200	770	0	0	—	—	—	199	0	.2	0	0	0	0	0
1 tablespoon	12	45	0	0	—	—	—	12	0	Trace	0	0	0	0	0
Lump, 1⅛ by ¾ by ⅜															
1 lump	6	25	0	0	—	—	—	6	0	Trace	0	0	0	0	0
Powdered, stirred before measuring.															
1 cup	128	495	0	0	—	—	—	127	0	.1	0	0	0	0	0
1 tablespoon	8	30	0	0	—	—	—	8	0	Trace	0	0	0	0	0
Brown, firm-packed															
1 cup	220	820	0	0	—	—	—	212	187	7.5	0	.02	.07	.4	0
1 tablespoon	14	50	0	0	—	—	—	13	12	.5	0	Trace	Trace	Trace	0
MISCELLANEOUS ITEMS															
Beer (average 3.6 percent alcohol by weight).															
1 cup	240	100	1	0	—	—	—	9	12	Trace	—	.01	.07	1.6	—
Beverages, carbonated:															
Cola type															
1 cup	240	95	0	0	—	—	—	24	—	—	0	0	0	0	0
Ginger ale															
1 cup	230	70	0	0	—	—	—	18	—	—	0	0	0	0	0
Bouillon cube, ⅝ inch															
1 cube	4	5	1	Trace	—	—	—	Trace	—	—	—	—	—	—	—

Food, Approximate Measure, and Weight (in Grams)		Food Energy	Protein	Fat (Total Lipid)	Fatty acids: Saturated (Total)	Fatty acids: Unsaturated Oleic	Fatty acids: Unsaturated Linoleic	Carbohydrate	Calcium	Iron	Vitamin A Value	Thiamine	Riboflavin	Niacin	Ascorbic Acid
	Grams	(Calories)	(Gm.)	(Gm.)	(Gm.)	(Gm.)	(Gm.)	(Gm.)	(Mo.)	(Mo.)	(I.U.)	(Mo.)	(Mo.)	(Mo.)	(Mo.)
Chili powder. See Vegetables, peppers.															
Chili sauce (mainly tomatoes).															
1 tablespoon	17	20	Trace	Trace	----	----	----	4	3	.1	240	.02	.01	.3	3
Chocolate:															
Bitter or baking															
1 ounce	28	145	3	15	8	6	Trace	8	22	1.9	20	.01	.07	.4	0
Sweet															
1 ounce	28	150	1	10	6	4	Trace	16	27	.4	Trace	.01	.04	.1	Trace
Cider. See Fruits, apple juice.															
Gelatin, dry:															
Plain															
1 tablespoon	10	35	9	Trace	----	----	----	----	----	----	----	----	----	----	----
Dessert powder, 3-ounce package.															
½ cup	85	315	8	0	----	----	----	75	----	----	----	----	----	----	----
Gelatin dessert, ready-to-eat:															
Plain															
1 cup	239	140	4	0	----	----	----	34	----	----	----	----	----	----	----
With fruit															
1 cup	241	160	3	Trace	----	----	----	40	----	----	----	----	----	----	----
Olives, pickled:															
Green															
4 medium or 3 extra large or 2 giant	16	15	Trace	2	Trace	2	Trace	Trace	8	.2	40	----	----	----	----
Ripe: Mission															
3 small or 2 large	10	15	Trace	2	Trace	2	Trace	Trace	9	.1	10	Trace	Trace	----	----
Pickles, cucumber:															
Dill, large, 4 by 1¾ inches.															
1 pickle	135	15	1	Trace	----	----	----	3	35	1.4	140	Trace	.03	Trace	8

Sweet, 2¾ by ¾ inches															
1 pickle	20	30	Trace	Trace	----	----	----	7	2	.2	20	Trace	Trace	Trace	1
Popcorn. See Grain products.															
Sherbet, orange															
1 cup	193	260	2	2	----	----	----	59	31	Trace	110	.02	.06	Trace	4
Soups, canned; ready-to-serve (prepared with equal volume of water):															
Bean with pork															
1 cup	250	170	8	6	1	2	2	22	62	2.2	650	.14	.07	1.0	2
Beef noodle															
1 cup	250	70	4	3	1	1	1	7	8	1.0	50	.05	.06	1.1	Trace
Beef bouillon, broth, consomme.															
1 cup	240	30	5	0	0	0	0	3	Trace	.5	Trace	Trace	.02	1.2	----
Chicken noodle															
1 cup	250	65	4	2	Trace	1	1	8	10	0.5	50	0.02	0.02	0.8	Trace
Clam chowder															
1 cup	255	85	2	3	----	----	----	13	36	1.0	920	.03	.03	1.0	----
Cream soup (mushroom)															
1 cup	240	135	2	10	1	3	5	10	41	.5	70	.02	.12	.7	Trace
Minestrone															
1 cup	245	105	5	3	----	----	----	14	37	1.0	2,350	.07	.05	1.0	----
Pea, green															
1 cup	245	130	6	2	1	1	Trace	23	44	1.0	340	.05	.05	1.0	7
Tomato															
1 cup	245	90	2	2	Trace	1	1	16	15	.7	1,000	.06	.05	1.1	12
Vegetable with beef broth.															
1 cup	250	80	3	2	----	----	----	14	20	.8	3,250	.05	.02	1.2	----
Starch (cornstarch)															
1 cup	128	465	Trace	Trace	----	----	----	112	0	0	0	0	0	0	0
1 tablespoon	8	30	Trace	Trace	----	----	----	7	0	0	0	0	0	0	0
Tapioca, quick-cooking granulated, dry, stirred before measuring.															
1 cup	152	535	1	Trace	----	----	----	131	15	.6	0	0	0	0	0
1 tablespoon	10	35	Trace	Trace	----	----	----	9	1	Trace	0	0	0	0	0
Vinegar															
1 tablespoon	15	2	0	----	----	----	----	1	1	.1	----	----	----	----	----
White sauce, medium															
1 cup	265	430	10	33	18	11	1	23	305	.5	1,220	.12	.44	.6	Trace

Food, Approximate Measure, and Weight (in Grams)		Food Energy	Protein	Fat (Total Lipid)	Fatty acids: Saturated (Total)	Fatty acids: Unsaturated: Oleic	Fatty acids: Unsaturated: Linoleic	Carbohydrate	Calcium	Iron	Vitamin A Value	Thiamine	Riboflavin	Niacin	Ascorbic Acid
	Grams	(Calories)	(Gm.)	(Gm.)	(Gm.)	(Gm.)	(Gm.)	(Gm.)	(Mo.)	(Mo.)	(I.U.)	(Mo.)	(Mo.)	(Mo.)	(Mo.)
Yeast:															
Compressed 1 ounce	28	25	3	Trace	----	----	----	3	4	1.4	Trace	.20	.47	3.2	Trace
Dry active 1 ounce	28	80	10	Trace	----	----	----	11	12	4.6	Trace	.66	1.53	10.4	Trace
Brewer's, dry, debittered. 1 tablespoon	8	25	3	Trace	----	----	----	3	17	1.4	Trace	1.25	.34	3.0	Trace
Yogurt. See Milk, cream, cheese; related products.															

AMINO ACID CONTENT OF FOODS, 100 GM., EDIBLE PORTION

Protein Content, and Nitrogen Conversion Factor	Tryptophan (Gm.)	Threonine (Gm.)	Isoleucine (Gm.)	Leucine (Gm.)	Lysine (Gm.)	Methionine (Gm.)	Cystine (Gm.)	Phenylalanine (Gm.)	Tyrosine (Gm.)	Valine (Gm.)	Arginine (Gm.)	Histidine (Gm.)
MILK; MILK PRODUCTS												
Milk (Ptn, N×6.38):												
cow:												
fluid, whole and non-fat (3.5% ptn)	0.049	0.161	0.223	0.344	0.272	0.086	0.031	0.170	0.178	0.240	0.128	0.092
canned:												
evaporated, unsweetened (7.0% ptn)	0.099	0.323	0.447	0.688	0.545	0.171	0.063	0.340	0.357	0.481	0.256	0.185
condensed, sweetened (8.1% ptn)	0.114	0.374	0.518	0.796	0.631	0.198	0.072	0.393	0.413	0.557	0.296	0.214
dried:												
whole (25.8% ptn)	0.364	1.191	1.648	2.535	2.009	0.632	0.231	1.251	1.316	1.774	0.944	0.680
non-fat (35.6% ptn)	0.502	1.641	2.271	3.493	2.768	0.870	0.318	1.724	1.814	2.444	1.300	0.937
human (1.4% ptn)	0.023	0.062	0.075	0.278	0.312	0.065	——	0.121	——	0.139	0.174	0.068
goat (3.3% ptn)	0.039	0.217	0.087	0.124	0.090	0.028	0.027	0.060	0.071	0.086	0.055	0.030
Indian buffalo (4.2% ptn)	0.059	0.212	0.204	0.420	0.331	0.112	0.058	0.177	——	0.255	0.136	0.086
Milk products:												
buttermilk (3.5% ptn, N×6.38)	0.038	0.165	0.219	0.348	0.291	0.082	0.032	0.186	0.137	0.262	0.168	0.099
casein (100.0% ptn, N×6.29)	1.335	4.277	6.550	10.048	8.013	3.084	0.382	5.389	5.819	7.393	4.070	3.021
cheese (ptn, N×6.38):												
blue mold (21.5% ptn)	0.293	0.799	1.449	2.096	1.577	0.559	0.121	1.153	1.028	1.543	0.785	0.701
Camembert (17.5% ptn)	0.239	0.650	1.179	1.706	1.284	0.455	0.099	0.938	0.837	1.256	0.639	0.571
Cheddar (25.0% ptn)	0.341	0.929	1.685	2.437	1.834	0.650	0.141	1.340	1.195	1.794	0.913	0.815
Cheddar processed (23.2% ptn)	0.316	0.862	1.563	2.262	1.702	0.604	0.131	1.244	1.109	1.665	0.847	0.756
cheese foods, Cheddar (20.5% ptn)	0.280	0.761	1.382	1.998	1.504	0.533	0.116	1.099	0.980	1.472	0.749	0.668
cottage (17.0% ptn)	0.179	0.794	0.989	1.826	1.428	0.469	0.147	0.917	0.917	0.978	0.802	0.549
cream cheese (9.0% ptn)	0.080	0.408	0.519	0.923	0.721	0.229	0.085	0.547	0.408	0.538	0.313	0.278
Limburger (21.2% ptn)	0.289	0.788	1.429	2.067	1.555	0.552	0.120	1.136	1.014	1.522	0.774	0.691
Parmesan (36.0% ptn)	0.491	1.337	2.426	3.510	2.641	0.937	0.203	1.930	1.721	2.584	1.315	1.174
Swiss (27.5% ptn)	0.375	1.021	1.853									
Swiss processed (26.4% ptn)	0.360	0.981	1.779	2.574	1.937	0.687	0.149	1.415	1.262	1.895	0.964	0.861
lactalbumin (100.0% ptn, N×6.49)	2.203	5.239	6.209	12.342	9.060	2.250	3.405	4.360	3.806	5.686	3.498	1.911
whey (Ptn, N×6.49)												
fluid (0.9% ptn)	0.010	0.048	0.052	0.074	0.055	0.013	0.018	0.023	0.009	0.045	0.017	0.011
dried (12.7% ptn)	0.147	0.677	0.734	1.043	0.769	0.188	0.250	0.323	0.131	0.640	0.235	0.159

Protein Content, and Nitrogen Conversion Factor	Tryptophan (Gm.)	Threonine (Gm.)	Isoleucine (Gm.)	Leucine (Gm.)	Lysine (Gm.)	Methionine (Gm.)	Cystine (Gm.)	Phenylalanine (Gm.)	Tyrosine (Gm.)	Valine (Gm.)	Arginine (Gm.)	Histidine (Gm.)
EGGS, CHICKEN (Ptn, N×6.25)												
Fresh or stored:												
whole (12.8% ptn)	0.211	0.637	0.850	1.126	0.819	0.401	0.299	0.739	0.551	0.950	0.840	0.307
whites (10.8% ptn)	0.164	0.477	0.698	0.950	0.648	0.420	0.263	0.689	0.449	0.842	0.634	0.233
yolks (16.3% ptn)	0.235	0.827	0.996	1.372	1.074	0.417	0.274	0.717	0.756	1.121	1.132	0.368
Dried:												
whole (46.8% ptn)	0.771	2.329	3.108	4.118	2.995	1.468	1.093	2.703	2.014	3.474	3.070	1.123
whites (85.9% ptn)	1.306	3.793	5.553	7.559	5.154	3.340	2.089	5.484	3.573	6.693	5.044	1.855
yolks (31.2% ptn)	0.449	1.582	1.907	2.626	2.057	0.799	0.524	1.373	1.448	2.147	2.167	0.704
MEAT; POULTRY; FISH AND SHELLFISH; THEIR PRODUCTS												
Meat (Ptn, N×6.25):												
beef carcass or side:												
thin (18.8% ptn)	0.220	0.830	0.984	1.540	1.642	0.466	0.238	0.773	0.638	1.044	1.212	0.653
medium fat (17.5% ptn)	0.204	0.773	0.916	1.434	1.529	0.434	0.221	0.720	0.594	0.972	1.128	0.608
fat (16.3% ptn)	0.190	0.720	0.853	1.335	1.424	0.404	0.206	0.670	0.553	0.905	1.051	0.566
very fat (13.7% ptn)	0.160	0.605	0.717	1.122	1.197	0.340	0.173	0.563	0.465	0.761	0.883	0.476
medium fat, trimmed to retail basis (18.2% ptn)	0.213	0.804	0.952	1.491	1.590	0.451	0.230	0.748	0.617	1.010	1.174	0.632
beef cuts, medium fat:												
chuck (18.6% ptn)	0.217	0.821	0.973	1.524	1.625	0.461	0.235	0.765	0.631	1.033	1.199	0.646
flank (19.9% ptn)	0.232	0.879	1.041	1.630	1.738	0.494	0.252	0.818	0.675	1.105	1.283	0.691
hamburger (16.0% ptn)	0.187	0.707	0.837	1.311	1.398	0.397	0.202	0.658	0.543	0.888	1.032	0.556
porterhouse (16.4% ptn)	0.192	0.724	0.858	1.343	1.433	0.407	0.207	0.674	0.556	0.911	1.057	0.569
rib roast (17.4% ptn)	0.203	0.768	0.910	1.425	1.520	0.432	0.220	0.715	0.590	0.966	1.122	0.604
round (19.5% ptn)	0.228	0.861	1.020	1.597	1.704	0.484	0.246	0.802	0.661	1.083	1.257	0.677
rump (16.2% ptn)	0.189	0.715	0.848	1.327	1.415	0.402	0.205	0.666	0.550	0.899	1.045	0.562
sirloin (17.3% ptn)	0.202	0.764	0.905	1.417	1.511	0.429	0.219	0.711	0.587	0.960	1.116	0.601
beef, canned (25.0% ptn)	0.292	1.104	1.308	2.048	2.184	0.620	0.316	1.028	0.848	1.388	1.612	0.868
beef, dried or chipped (34.3% ptn)	0.401	1.515	1.795	2.810	2.996	0.851	0.434	1.410	1.163	1.904	2.212	1.191
lamb carcass or side:												
thin (17.1% ptn)	0.222	0.782	0.886	1.324	1.384	0.410	0.224	0.695	0.594	0.843	1.114	0.476
medium fat (15.7% ptn)	0.203	0.718	0.814	1.216	1.271	0.377	0.206	0.638	0.545	0.774	1.022	0.437
fat (13.0% ptn)	0.168	0.595	0.674	1.007	1.052	0.312	0.171	0.528	0.451	0.641	0.847	0.362

MEAT; POULTRY; FISH—Continued

lamb cuts, medium fat:												
leg (18.0% ptn)	0.233	0.824	0.933	1.394	1.457	0.432	0.236	0.732	0.625	0.887	1.172	0.501
rib (14.9% ptn)	0.193	0.682	0.772	1.154	1.206	0.358	0.195	0.606	0.517	0.734	0.970	0.415
shoulder (15.6% ptn)	0.202	0.714	0.809	1.208	1.263	0.374	0.205	0.634	0.542	0.769	1.016	0.434
pork, packer's carcass or side:												
thin (14.1% ptn)	0.183	0.654	0.724	1.038	1.157	0.352	0.165	0.555	0.503	0.733	0.864	0.487
medium fat (11.9% ptn)	0.154	0.552	0.611	0.876	0.977	0.297	0.139	0.468	0.425	0.619	0.729	0.411
fat (9.8% ptn)	0.127	0.455	0.503	0.721	0.804	0.245	0.114	0.386	0.350	0.510	0.601	0.339
pork cuts, medium fat, fresh:												
ham (15.2% ptn)	0.197	0.705	0.781	1.119	1.248	0.379	0.178	0.598	0.542	0.790	0.931	0.525
loin (16.4% ptn)	0.213	0.761	0.842	1.207	1.346	0.409	0.192	0.646	0.585	0.853	1.005	0.567
miscellaneous lean cuts (14.5% ptn)	0.188	0.673	0.745	1.067	1.190	0.362	0.169	0.571	0.517	0.754	0.889	0.501
pork, cured:												
bacon, medium fat (9.1% ptn)	0.095	0.306	0.399	0.728	0.587	0.141	0.106	0.434	0.234	0.434	0.622	0.246
fat back or salt pork (3.9% ptn)	0.006	0.141	0.110	0.367	0.317	0.055	0.043	0.157	0.052	0.168	0.379	0.035
ham (16.9% ptn)	0.162	0.162	0.841	1.306	1.420	0.411	0.273	0.646	0.652	0.879	1.068	0.544
luncheon meat:												
boiled ham (22.8% ptn)	0.219	0.934	1.135	1.762	1.915	0.554	0.368	0.872	0.879	1.186	1.441	0.733
canned, spiced (14.9% ptn)	0.143	0.610	0.741	1.151	1.252	0.362	0.241	0.570	0.575	0.775	0.942	0.479
rabbit, domesticated, flesh only (21.0% ptn)	—	1.021	1.082	1.636	1.818	0.541	—	0.793	—	1.021	1.176	0.474
veal, carcass or side:												
thin (19.7% ptn)	0.258	0.854	1.040	1.444	1.645	0.451	0.233	0.801	0.709	1.018	1.283	0.634
medium fat (19.1% ptn)	0.251	0.828	1.008	1.400	1.595	0.437	0.226	0.776	0.688	0.987	1.244	0.614
fat (18.5% ptn)	0.243	0.802	0.977	1.356	1.545	0.423	0.219	0.752	0.666	0.956	1.205	0.595
veal cuts, medium fat:												
round (19.5% ptn)	0.256	0.846	1.030	1.429	1.629	0.446	0.231	0.792	0.702	1.008	1.270	0.627
shoulder (19.4% ptn)	0.255	0.841	1.024	1.422	1.620	0.444	0.230	0.788	0.698	1.003	1.263	0.624
stew meat (18.3% ptn)	0.240	0.793	0.966	1.341	1.528	0.419	0.217	0.744	0.659	0.946	1.192	0.589
Poultry (Ptn, N×6.25):												
chicken, flesh only:												
broilers or fryers (20.6% ptn)	0.250	0.877	1.088	1.490	1.810	0.537	0.277	0.811	0.725	1.012	1.302	0.593
hens (21.3% ptn)	0.259	0.907	1.125	1.540	1.871	0.556	0.286	0.838	0.750	1.046	1.346	0.613
ducks, domesticated, flesh only (21.4% ptn)	—	0.935	1.109	1.657	1.842	0.531	—	0.842	—	1.027	1.301	0.486
turkey, flesh only (24.0% ptn)	—	1.014	1.260	1.836	2.173	0.664	0.330	0.960	—	1.187	1.513	0.649

Protein Content, and Nitrogen Conversion Factor	Tryptophan (Gm.)	Threonine (Gm.)	Isoleucine (Gm.)	Leucine (Gm.)	Lysine (Gm.)	Methionine (Gm.)	Cystine (Gm.)	Phenylalanine (Gm.)	Tyrosine (Gm.)	Valine (Gm.)	Arginine (Gm.)	Histidine (Gm.)
MEAT; POULTRY; FISH—Continued												
Products from meat, poultry, and fish (Ptn, N×6.25):												
brains (10.4% ptn)	0.138	0.494	0.504	0.845	0.760	0.220	0.145	0.506	0.433	0.536	0.614	0.278
chitterlings (8.6% ptn)	0.094	0.398	0.308	0.457	0.670	0.193	0.109	0.359	0.228	0.462	1.406	0.169
fish flour (76.0% ptn)	0.754	4.378	4.232	6.189	7.381	2.019	----	2.845	----	3.916	5.204	1.289
gelatin (85.6% ptn, N×5.55)	0.006	1.912	1.357	2.930	4.226	0.787	0.077	2.036	0.401	2.421	7.866	0.771
gizzard, chicken (23.1% ptn)	0.207	1.072	1.094	1.689	1.567	0.554	0.218	0.968	0.680	1.116	1.741	0.480
heart:												
beef or pork (16.9% ptn)	0.219	0.776	0.857	1.509	1.387	0.403	0.168	0.765	0.627	0.973	1.068	0.433
chicken (20.5% ptn)	0.266	0.941	1.040	1.830	1.683	0.489	0.203	0.928	0.761	1.181	1.296	0.525
kidney:												
beef (15.0% ptn)	0.221	0.665	0.730	1.301	1.087	0.307	0.182	0.706	0.557	0.876	0.934	0.377
pork (16.3% ptn)	0.240	0.722	0.793	1.414	1.181	0.334	0.198	0.767	0.605	0.952	1.015	0.409
sheep (16.6% ptn)	0.244	0.736	0.807	1.440	1.203	0.340	0.202	0.781	0.616	0.969	1.033	0.417
liver:												
beef or pork (19.7% ptn)	0.296	0.936	1.031	1.819	1.475	0.463	0.243	0.993	0.738	1.239	1.201	0.523
calf (19.0% ptn)	0.286	0.903	0.994	1.574	1.423	0.447	0.234	0.958	0.711	1.195	1.158	0.505
chicken (22.1% ptn)	0.332	1.050	1.156	2.040	1.655	0.520	0.272	1.114	0.827	1.390	1.347	0.587
sheep or lamb (21.0% ptn)	0.316	0.998	1.099	1.939	1.572	0.494	0.259	1.058	0.786	1.320	1.280	0.558
pancreas:												
beef (13.5% ptn)	0.175	0.626	0.683	1.054	0.996	0.244	----	0.562	0.590	0.724	0.771	0.266
pork (14.5% ptn)	0.188	0.673	0.733	1.132	1.070	0.262	----	0.603	0.633	0.777	0.828	0.285
pork or beef, canned (14.3% ptn)	0.151	0.618	0.730	1.190	1.345	0.327	0.261	0.579	0.570	0.810	1.050	0.460
potted meat (16.1% ptn)	0.149	0.662	0.641	1.203	1.061	0.361	----	0.641	----	0.943	1.002	0.322
sausage:												
Bologna (14.8% ptn)	0.126	0.606	0.718	1.061	1.191	0.313	0.185	0.540	0.481	0.744	1.028	0.308
Braunschweiger (15.4% ptn)	0.172	0.668	0.754	1.291	1.200	0.320	0.187	0.700	0.471	0.956	0.954	0.458
frankfurters (14.2% ptn)	0.120	0.582	0.688	1.018	1.143	0.300	0.177	0.518	0.461	0.713	0.986	0.382
head cheese (15.0% ptn)	0.079	0.418	0.509	0.946	0.907	0.250	0.209	0.569	0.569	0.617	1.075	0.278
liverwurst (16.7% ptn)	0.187	0.724	0.818	1.400	1.301	0.347	0.203	0.759	0.510	1.037	1.034	0.497
pork, links or bulk, raw (10.8% ptn)	0.092	0.442	0.524	0.774	0.869	0.228	0.135	0.394	0.351	0.543	0.750	0.290
pork, bulk, canned (15.4% ptn)	0.131	0.631	0.747	1.104	1.239	0.325	0.192	0.562	0.500	0.774	1.069	0.414
salami (23.9% ptn)	0.203	0.979	1.159	1.713	1.923	0.505	0.298	0.872	0.776	1.201	1.660	0.642

MEAT; POULTRY; FISH—Continued

Vienna sausage, canned (15.8% ptn)	0.134	0.647	0.766	1.133	1.272	0.334	0.197	0.576	0.513	0.794	1.097	0.425
tongue:												
beef (16.4% ptn)	0.197	0.708	0.792	1.286	1.364	0.357	0.207	0.661	0.548	0.840	1.065	0.412
pork (16.8% ptn)	0.202	0.726	0.812	1.317	1.398	0.366	0.212	0.677	0.562	0.860	1.091	0.422
veal and pork loaf, canned (17.2% ptn)	0.198	0.627	0.859	1.236	1.258	0.418	0.209	0.619	0.468	0.958	0.916	0.388
LEGUMES (DRY SEED); COMMON NUTS; OTHER NUTS AND DRY SEEDS; THEIR PRODUCTS:												
Legume seeds and their products:												
beans (Phaseolus vulgaris) (N×6.25):												
pinto and red Mexican (23.0% ptn)	0.213	0.997	1.306	1.976	1.708	0.232	0.228	1.270	0.887	1.395	1.384	0.655
red kidney:												
raw (23.1% ptn)	0.214	1.002	1.312	1.985	1.715	0.233	0.229	1.275	0.891	1.401	1.390	0.658
canned, solids and liquid (5.7% ptn)	0.053	0.247	0.324	0.490	0.423	0.057	0.057	0.315	0.220	0.346	0.343	0.162
other common beans including navy, peabean, white marrow:												
raw (21.4% ptn)	0.199	0.928	1.216	1.839	1.589	0.216	0.212	1.181	0.825	1.298	1.287	0.609
baked with pork, canned (5.8% ptn)	0.057	0.274	0.291	0.486	0.354	0.059	0.018	0.333	0.165	0.312	0.251	0.186
black gram, raw (23.6% ptn, N×6.25)	0.242	0.801	1.390	2,062	1.510	0.332	0.287	1.242	0.551	1.450	1.552	0.559
broadbeans, raw (25.4% ptn, N×6.25)	0.236	0.829	1.593	2.211	1.426	0.106	0.179	1.057	0.687	1.276	1.780	0.748
chickpeas (20.8% ptn, N±6.25)	0.170	0.739	1.195	1.538	1.434	0.276	0.296	1.012	0.692	1.025	1.551	0.559
cowpeas (22.9% ptn, N×6.25)	0.220	0.901	1.110	1.715	1.491	0.352	0.297	1.198	0.678	1.293	1.473	0.692
dolichos, twin flower (21.6% ptn, N×6.25)	0.221	0.836	1.448	1.707	1.700	0.294	0.480	1.486	0.560	1.286	1.230	0.650
lentils, whole (25.0% ptn, N×6.25)	0.216	0.896	1.316	1.760	1.528	0.180	0.204	1.104	0.664	1.360	1.908	0.548
lima beans (20.7% ptn, N×6.25)	0.195	0.980	1.199	1.722	1.378	0.331	0.311	1.222	0.543	1.298	1.315	0.669
lupine (32.3% ptn, N×6.25)		1.101	1.618	1.964	1.447	0.114	----	1.271	----	1.328	2.718	0.811
moth beans (24.4% ptn, N×6.25)	0.164	----	1.093	1.484	1.202	0.191	0.109	1.003	1.245	0.695	----	0.722
mung beans (24.4% ptn, 6.25)	0.180	0.765	1.351	2.202	1.667	0.265	0.152	1.167	0.390	1.444	1.370	0.543
peanuts (26.9% ptn, N×5.46)	0.340	0.828	1.266	1.872	1.099	0.271	0.463	1.557	1.104	1.532	3.296	0.749
peanut flour (51.2% ptn, N×5.46)	0.647	1.575	2.410	3.563	2.091	0.516	0.881	2.963	2.100	2.916	6.273	1.425
peanut butter (26.1% ptn, N×5.46)	0.330	0.803	1.228	1.816	1.066	0.263	0.449	1.510	1.071	1.487	3.198	0.727
peas (Pisum salivum) (N×6.25):												
entire seeds (23.8% ptn)	0.251	0.918	1.340	1.969	1.744	0.286	0.308	1.200	0.960	1.333	2.102	0.651
split (24.5% ptn)	0.259	0.945	1.380	2.027	1.795	0.294	0.318	1.235	0.988	1.372	2.164	0.670
pigeonpeas, without seed coat (21.9% ptn, N×6.25)	0.119	0.834	1.346	1.717	1.580	0.256	0.308	1.875	0.725	1.153	1.489	0.617

Protein Content, and Nitrogen Conversion Factor	Tryptophan (Gm.)	Threonine (Gm.)	Isoleucine (Gm.)	Leucine (Gm.)	Lysine (Gm.)	Methionine (Gm.)	Cystine (Gm.)	Phenylalanine (Gm.)	Tyrosine (Gm.)	Valine (Gm.)	Arginine (Gm.)	Histidine (Gm.)
LEGUMES; SEEDS; NUTS—Continued												
soybeans, whole (34.9% ptn, N×5.71)	0.526	1.504	2.054	2.946	2.414	0.513	0.678	1.889	1.216	2.005	2.763	0.911
soybean flour, flakes, and grits (ptn, N×5.71):												
low fat (44.7% ptn)	0.673	1.926	2.630	3.773	3.092	0.658	0.869	2.419	1.558	2.568	3.538	1.166
medium fat (42.5% ptn)	0.640	1.831	2.501	3.588	2.940	0.625	0.826	2.300	1.481	2.441	3.364	1.109
full fat (35.9% ptn)	0.541	1.547	2.112	3.030	2.483	0.528	0.698	1.943	1.251	2.062	2.842	0.937
soybean curd (7.0% ptn, N×5.71)	----	----	----	----	----	0.081	0.091	----	----	----	----	----
soybean milk (3.4% ptn, N×5.71)	0.051	0.176	0.175	0.305	0.269	0.054	0.071	0.195	0.193	0.186	0.302	0.121
vetch (28.8% ptn, N×6.25)	0.203	0.899	2.198	2.290	1.898	0.346	0.336	1.014	0.369	1.442	2.249	0.659
Common nuts and their products:												
almonds (18.6% ptn, N×5.18)	0.176	0.610	0.873	1.454	0.582	0.259	0.377	1.146	0.618	1.124	2.729	0.517
Brazil nuts (14.4% ptn, N×5.46)	0.187	0.422	0.593	1.129	0.443	0.941	0.504	0.617	0.483	0.823	2.247	0.367
cashews (18.5% ptn, N×5.30)	0.471	0.737	1.222	1.522	0.792	0.353	0.527	0.946	0.712	1.592	2.098	0.415
coconut (3.4% ptn, N×5.30)	0.033	0.129	0.180	0.269	0.152	0.071	0.062	0.174	0.101	0.212	0.486	0.069
coconut meal (20.3% ptn, N×5.30)	0.199	0.770	1.076	1.605	0.908	0.421	0.372	1.038	0.605	1.268	2.899	0.414
filberts (12.7% ptn, N×5.30)	0.211	0.415	0.853	0.939	0.417	0.139	0.165	0.537	0.434	0.934	2.171	0.288
peanuts. See Legumes.												
pecans (9.4% ptn, N×5.30)	0.138	0.389	0.553	0.773	0.435	0.153	0.216	0.564	0.316	0.525	1.185	0.273
walnuts (English or Persian) (15.0% ptn, N×5.30)	0.175	0.589	0.767	1.228	0.441	0.306	0.320	0.767	0.583	0.974	2.287	0.405
Other nuts and seeds and their products												
acorns (10.4% ptn)	0.126	0.434	0.561	0.808	0.636	0.139	0.184	0.473	----	0.718	0.722	0.251
amaranth (14.6% ptn)	0.149	0.832	0.882	1.209	1.074	0.372	0.521	1.141	----	0.849	1.747	0.441
balsam pear seed meal (41.9% ptn)	----	----	----	----	1.265	----	0.142	2.609	0.617	----	5.914	0.917
breadnut tree, Ramon (9.6% ptn)	0.261	0.373	0.543	1.041	0.418	0.056	----	0.453	----	0.927	0.884	0.147
Chinese tallow tree-nut flour (57.6% ptn)	0.837	2.174	3.510	4.347	1.587	0.924	0.696	2.847	2.011	4.510	10.031	1.587
chocolate tree, Nicaragua (38.5% ptn)	0.588	1.496	2.092	3.952	2.223	0.276	----	2.630	----	2.404	4.220	0.683
cottonseed flour and meal (42.3% ptn)	0.591	1.764	1.884	2.945	2.139	0.686	0.814	2.610	1.365	2.458	5.603	1.325
earpod tree, Guanacaste (34.1% ptn)	0.444	1.165	2.213	4.581	1.930	0.360	----	1.325	----	1.570	2.857	1.004
lead tree (24.1% ptn)	0.191	0.828	1.651	1.787	1.164	0.055	----	0.855	----	0.864	2.410	0.565
pumpkin seed (30.9% ptn)	0.560	0.933	1.737	2.437	1.411	0.577	----	1.749	----	1.679	4.810	0.711

LEGUMES; SEEDS; NUTS—Continued

safflower seed meal (42.1% ptn)	0.675	1.462	1.914	2.740	1.525	0.731	——	2.605	——	2.446	4.623	0.985
sesame:												
seed (19.3% ptn)	0.331	0.707	0.951	1.679	0.583	0.637	0.495	1.457	0.951	0.885	1.992	0.441
meal (33.4% ptn)	0.573	1.223	1.645	2.905	1.008	1.103	0.857	2.521	1.645	1.531	3.447	0.763
sunflower:												
kernel (23.0% ptn)	0.343	0.911	1.276	1.736	0.868	0.443	0.464	1.220	0.640	1.354	2.370	0.586
meal (39.5% ptn)	0.589	1.565	2.191	2.981	1.491	0.760	0.797	2.094	1.110	2.325	4.069	1.006
GRAINS AND THEIR PRODUCTS												
Barley (12.8% ptn, N×5.83)	0.160	0.433	0.545	0.889	0.433	0.184	0.257	0.661	0.466	0.643	0.659	0.239
Bread, white (4% non-fat dry milk, flour basis) (8.5% ptn, N×5.70)	0.091	0.282	0.429	0.668	0.225	0.142	0.200	0.465	0.243	0.435	0.340	0.192
Buckwheat flour:												
dark (11.7% ptn, N×6.25)	0.165	0.461	0.440	0.683	0.687	0.206	0.228	0.442	0.240	0.607	0.930	0.256
light (6.4% ptn, N×6.25)	0.090	0.252	0.241	0.374	0.376	0.113	0.125	0.242	0.131	0.332	0.509	0.140
Canihua (14.7% ptn, N×6.25)	0.118	0.706	1.000	0.851	0.882	0.263	0.162	0.529	0.294	0.677	1.162	0.367
Cereal combinations:												
corn and soy grits (18.0% ptn, N×6.25)	0.116	0.792	0.841	1.656	0.772	0.271	0.311	0.832	0.562	1.054	0.982	0.472
infant food, precooked, mixed cereals with non-fat dry milk and yeast (19.4% ptn, N×6.25)	0.118	——	——	——	0.273	0.310	0.137	0.543	0.447	——	0.447	0.233
oat-corn-rye mixture, puffed (14.5% ptn, N×5.83)	0.172	0.545	0.841	1.368	0.343	0.388	0.234	0.933	0.622	0.900	0.776	0.326
Corn, field (10.0% ptn, N×6.25)	0.061	0.398	0.462	1.296	0.288	0.186	0.130	0.454	0.611	0.510	0.352	0.206
Corn flour (7.8% ptn, N×6.25)	0.047	0.311	0.361	1.011	0.225	0.145	0.101	0.354	0.477	0.398	0.275	0.161
Corn grits (8.7% ptn, N×6.25)	0.053	0.347	0.402	1.128	0.251	0.161	0.113	0.395	0.532	0.444	0.306	0.180
Cornmeal:												
whole ground (9.2% ptn, N×6.25)	0.056	0.367	0.425	1.192	0.265	0.171	0.119	0.418	0.562	0.470	0.324	0.190
degermed (7.9% ptn, N×6.25)	0.048	0.315	0.365	1.024	0.228	0.147	0.102	0.359	0.483	0.403	0.278	0.163
Corn products:												
flakes (8.1% ptn, N×6.25)	0.052	0.275	0.306	1.047	0.154	0.135	0.152	0.354	0.283	0.386	0.231	0.226
germ (1.4% ptn, N×6.25)	0.144	0.622	0.578	1.030	0.791	0.232	0.130	0.483	0.343	0.789	1.134	0.464
gluten (10.0% ptn, N×6.25)	0.059	0.344	0.443	1.563	0.179	0.282	0.141	0.558	0.582	0.512	0.322	0.200
hominy (8.7% ptn, N×6.25)	0.084	0.316	0.349	0.810	0.358	0.099	——	0.333	0.331	0.398	0.444	0.203
masa (2.8% ptn, N×6.25)	0.010	——	——	——	0.103	0.108	0.030	——	——	——	——	——
pozol (5.9% ptn, N×6.25)	0.042	0.336	0.304	0.591	0.234	0.087	——	0.254	——	0.267	0.197	0.122
tortilla (5.8% ptn, N×6.25)	0.031	0.235	0.345	0.939	0.145	0.111	——	0.252	——	0.304	0.223	0.128

Protein Content, and Nitrogen Conversion Factor	Tryptophan (Gm.)	Threonine (Gm.)	Isoleucine (Gm.)	Leucine (Gm.)	Lysine (Gm.)	Methionine (Gm.)	Cystine (Gm.)	Phenylalanine (Gm.)	Tyrosine (Gm.)	Valine (Gm.)	Arginine (Gm.)	Histidine (Gm.)
GRAINS AND PRODUCTS—Contiued												
zein (16.1% ptn, N×6.25)	0.010	0.495	0.822	3.184	----	0.281	0.162	1.664	0.981	0.654	0.286	0.216
Job's tears (13.8% ptn, N×5.83)	0.066	0.620	1.065	3.506	0.362	0.459	0.265	0.703	----	----	0.518	0.317
Millets:												
foxtail millet (9.7% ptn, N×5.83)	0.103	0.323	0.790	1.737	0.218	0.291	----	0.697	----	0.717	0.374	0.218
little millet (7.2% ptn, N×5.83)	0.047	0.262	0.517	0.841	0.188	0.178	----	0.370	----	0.471	0.363	0.147
pearl millet (11.4% ptn, N×5.83)	0.248	0.456	0.635	1.746	0.383	0.270	0.152	0.506	----	0.682	0.524	0.240
ragimillet (6.2% ptn, N×5.83)	0.085	0.270	0.398	0.620	0.202	0.270	0.187	0.263	----	0.473	0.100	0.079
Oatmeal and rolled oats (14.2% ptn, N×5.83)	0.183	0.470	0.733	1.065	0.521	0.209	0.309	0.758	0.524	0.845	0.935	0.261
Quinoa (11.0% ptn, N×6.25)	0.120	0.523	0.722	0.781	0.729	0.278	0.107	0.394	0.253	0.447	0.820	0.297
Rice:												
brown (7.5% ptn, N×5.95)	0.081	0.294	0.352	0.646	0.296	0.135	0.102	0.377	0.343	0.524	0.432	0.126
white and converted (7.6% ptn, N×5.95)	0.082	0.298	0.356	0.655	0.300	0.137	0.103	0.382	0.347	0.531	0.438	0.128
?ice products:												
flakes or puffed (5.9% ptn, N×5.95)	0.046	----	----	----	0.056	----	0.044	0.286	0.124	----	0.137	0.137
germ (14.2% ptn, N×5.95)	0.270	2.177	0.630	0.838	1.707	0.420	0.169	0.750	0.929	0.938	1.559	0.430
Rye (12.1% ptn, N×5.83)	0.137	0.448	0.515	0.813	0.494	0.191	0.241	0.571	0.390	0.631	0.591	0.276
Rye flour:												
light (9.4% ptn, N×5.83)	0.106	0.348	0.400	0.632	0.384	0.148	0.187	0.443	0.303	0.490	0.459	0.214
medium (11.4% ptn, N×5.83)	0.129	0.422	0.485	0.766	0.465	0.180	0.227	0.538	0.368	0.594	0.557	0.260
Sorghum (11.0% ptn, N×6.25)	0.123	0.394	0.598	1.767	0.299	0.190	0.183	0.547	0.303	0.628	0.417	0.211
Teosinte (22.0% ptn, N×6.25)	0.049	----	----	----	0.348	0.496	----	----	----	----	----	----
Wheat, whole grain:												
hard red spring (14.0% ptn, N×5.83)	0.173	0.403	0.607	0.939	0.384	0.214	0.307	0.691	0.523	0.648	0.670	0.286
hard red winter (12.3% ptn, N×5.83)	0.152	0.354	0.534	0.825	0.338	0.188	0.270	0.608	0.460	0.570	0.589	0.251
soft red winter (10.2% ptn, N×5.83)	0.126	0.294	0.443	0.684	0.280	0.156	0.224	0.504	0.382	0.472	0.488	0.208
white (9.4% ptn, N×5.83)	0.116	0.271	0.408	0.630	0.258	0.143	0.206	0.464	0.351	0.435	0.450	0.192
durum (12.7% ptn, N×5.83)	0.157	0.306	0.551	0.852	0.348	0.194	0.279	0.627	0.475	0.588	0.608	0.259
Wheat flour:												
whole grain (13.3% ptn, N×5.83)	0.164	0.383	0.577	0.892	0.365	0.203	0.292	0.657	0.497	0.616	0.636	0.271
Intermediate extraction (12.0% ptn, N×5.70)	----	0.392	0.619	0.924	0.356	0.198	0.320	0.732	0.335	0.583	0.549	0.286
white (10.5% ptn, N×5.70)	0.129	0.302	0.483	0.809	0.239	0.138	0.210	0.577	0.359	0.453	0.466	0.210

Wheat products:												
bran (12.0% ptn, N×6.31)	0.196	0.342	0.484	0.717	0.491	0.145	0.270	0.434	0.259	0.552	0.742	0.280
bulgur (12.4% ptn, N×5.83)	0.070	——	——	——	0.430	0.300	0.319	——	——	——	——	——
farina (10.9% ptn, N×5.70)	0.124	——	——	——	0.199	0.143	0.184	0.579	0.447	——	0.424	0.268
flakes (10.8% ptn, N×5.70)	0.121	0.356	0.496	0.891	0.360	0.127	0.191	0.478	0.311	0.572	0.559	0.231
germ (25.2% ptn, N×5.80)	0.265	1.343	1.177	1.708	1.534	0.404	0.287	0.908	0.882	1.364	1.825	0.687
gluten, commercial (80.0% ptn, N×5.70)	0.856	2.119	3.677	5.993	1.530	1.389	1.726	4.351	2.596	3.789	3.481	1.825
gluten flour (41.4% ptn, N×5.70)	0.443	1.097	1.903	3.101	0.792	0.719	0.893	2.252	1.344	1.961	1.801	0.944
macaroni or spaghetti (12.8% ptn, N× 5.70	0.150	0.499	0.642	0.849	0.413	0.193	0.243	0.669	0.422	0.728	0.582	0.303
noodles, containing egg solids (12.6% ptn, N×5.70)	0.133	0.533	0.621	0.834	0.411	0.212	0.245	0.610	0.312	0.745	0.621	0.301
Shredded Wheat (10.1% ptn, N×5.83)	0.085	0.405	0.449	0.684	0.331	0.139	0.204	0.481	0.236	0.577	0.523	0.236
whole wheat with added germ (12.8% ptn, N×5.83)	0.136	——	——	——	0.466	——	0.246	0.755	0.481	——	0.742	0.731
FRUITS (PROTEIN, N×6.25)												
Abiu (1.7% ptn)	0.028	——	——	——	0.085	0.013	——	——	——	——	——	——
Avocados (1.3% ptn)	0.014	——	——	——	0.074	0.012	——	——	——	——	——	——
Bananas ripe												
common (1.2% ptn)	0.018	——	——	——	0.055	0.011	——	——	0.031	——	——	——
dwarf (1.2% ptn)	0.012	——	——	——	0.049	0.004	——	——	——	——	——	——
Dates (2.2% ptn)	0.061	0.061	0.074	0.077	0.065	0.027	——	0.063	——	0.094	0.049	0.049
Grapefruit (0.5% ptn)	0.001	——	——	——	0.006	0.000	——	——	——	——	——	——
Guavas, common (1.0% ptn)	0.010	——	——	——	0.030	0.010	——	——	——	——	——	——
Limes (0.8% ptn)	0.003	——	——	——	0.015	0.002	——	——	——	——	——	——
Mamey (0.5% ptn)	0.006	——	——	——	0.040	0.007	——	——	——	——	——	——
Mangos (0.7% ptn)	0.014	——	——	——	0.093	0.008	——	——	——	——	——	——
Muskmelons (0.6% ptn)	0.001	——	——	——	0.015	0.002	——	——	——	——	——	——
Oranges, sweet (0.9% ptn)	0.003	——	——	——	0.024	0.003	——	——	——	——	——	——
Orange juice (0.8% ptn)	0.003	——	——	——	0.021	0.002	——	——	——	——	——	——
Oranges, mandarin including tangerines (0.8% ptn)	0.005	——	——	——	0.028	0.004	——	——	——	——	——	——
Papayas (0.6% ptn)	0.012	——	——	——	0.038	0.002	——	——	——	——	——	——
Pineapple (0.4% ptn)	0.005	——	——	——	0.009	0.001	——	——	——	——	——	——
Plantain or baking banana (1.1% ptn)	0.010	0.027	0.056	0.059	0.050	0.005	0.016	0.049	——	0.065	0.045	——

Protein Content, and Nitrogen Conversion Factor	Tryptophan (Gm.)	Threonine (Gm.)	Isoleucine (Gm.)	Leucine (Gm.)	Lysine (Gm.)	Methionine (Gm.)	Cystine (Gm.)	Phenylalanine (Gm.)	Tyrosine (Gm.)	Valine (Gm.)	Arginine (Gm.)	Histidine (Gm.)
Soursop (1.0% ptn)	0.011	—	—	—	0.060	0.007	—	—	—	—	—	—
Sugarapple (1.8% ptn)	0.009	—	—	—	0.071	0.008	—	—	—	—	—	—
VEGETABLES												
Immature seeds (Ptn, N×6.25):												
corn, sweet, white or yellow:												
raw (3.7% ptn)	0.023	0.151	0.137	0.407	0.137	0.072	0.062	0.207	0.124	0.231	0.174	0.095
canned, solids and liquid (2.0% ptn)	0.012	0.082	0.074	0.220	0.074	0.039	0.033	0.112	0.067	0.125	0.094	0.052
cowpeas (9.4% ptn)	0.099	0.353	0.465	0.653	0.617	0.131	—	0.523	—	0.513	0.615	0.310
lima beans:												
raw (7.5% ptn)	0.097	0.388	0.460	0.605	0.474	0.080	0.083	0.389	0.259	0.485	0.454	0.247
canned, solids and liquid (3.8% ptn)	0.049	0.171	0.233	0.306	0.240	0.041	0.042	0.197	0.131	0.246	0.230	0.125
peas:												
raw (6.7% ptn)	0.056	0.245	0.308	0.418	0.316	0.054	0.073	0.257	0.163	0.274	0.595	0.109
canned, solids and liquid (3.4% ptn)	0.028	0.125	0.156	0.212	0.160	0.027	0.037	0.131	0.083	0.139	0.302	0.055
Leafy vegetables, raw (Ptn, N×6.25):												
amaranth (3.5% ptn)	0.038	0.056	0.164	0.206	0.141	0.025	0.024	0.096	0.105	0.136	0.134	0.069
beet greens (2.0% ptn)	0.024	0.076	0.084	0.129	0.108	0.034	—	0.116	—	0.101	0.083	0.026
Brussels sprouts (4.4% ptn)	0.044	0.153	0.185	0.194	0.197	0.046	—	0.148	—	0.193	0.279	0.106
cabbage (1.4% ptn)	0.011	0.039	0.040	0.057	0.066	0.013	0.028	0.030	0.030	0.043	0.105	0.025
chard (1.4% ptn)	0.014	0.058	0.060	0.076	0.055	0.004	—	0.046	—	0.055	0.035	0.018
chicory (1.6% ptn)	0.024	—	—	—	0.052	0.016	0.006	—	0.040	—	—	0.024
collards (3.9% ptn)	0.055	0.114	0.121	0.218	0.202	0.046	0.059	0.124	0.151	0.195	0.258	0.087
kale (3.9% ptn)	0.042	0.139	0.133	0.252	0.121	0.035	0.036	0.158	—	0.184	0.202	0.062
lettuce (1.2% ptn)	0.012	—	—	—	0.070	0.004	—	—	—	—	—	—
mustard greens (2.3% ptn)	0.037	0.060	0.075	0.062	0.111	0.024	0.035	0.074	0.121	0.108	0.167	0.041
parsley, curly garden (2.5% ptn)	0.050	—	—	—	0.160	0.012	—	—	—	—	—	—
spinach (2.3% ptn)	0.037	0.102	0.107	0.176	0.142	0.039	0.046	0.099	0.073	0.126	0.116	0.049
turnip greens (2.9% ptn)	0.045	0.125	0.107	0.207	0.129	0.052	0.045	0.146	0.105	0.149	0.167	0.051
watercress (1.7% ptn)	0.028	0.884	0.076	0.131	0.091	0.010	—	0.062	0.036	0.084	0.053	0.034
Starchy roots and tubers (Ptn, N×6.25):												
Apio arracacia (1.2% ptn)	0.008	—	—	—	0.042	0.003	—	—	—	—	—	—
cassava:												

flour (1.6% ptn)	0.021	0.044	0.045	0.066	0.066	0.010	0.018	0.045	0.030	0.049	0.159	0.025
root (1.1% ptn)	0.014	0.030	0.031	0.045	0.045	0.007	0.012	0.031	0.021	0.033	0.110	0.017
potatoes:												
raw (2.0% ptn)	0.021	0.079	0.088	0.100	0.107	0.025	0.019	0.088	0.036	0.107	0.099	0.029
canned, solids and liquid (1.7% ptn)	0.018	0.067	0.075	0.085	0.091	0.021	0.016	0.075	0.030	0.091	0.084	0.024
flour (7.1% ptn)	0.076	0.279	0.311	0.353	0.378	0.089	0.068	0.314	0.127	0.379	0.350	0.102
sweet potatoes (Ipomaea balatas):												
raw (1.8% ptn)	0.031	0.085	0.087	0.103	0.085	0.033	0.029	0.100	0.081	0.135	0.094	0.036
dehydrated (5.0% ptn)	0.087	0.235	0.241	0.286	0.236	0.093	0.080	0.278	0.225	0.374	0.261	0.099
taro (1.9% ptn)	0.035	0.089	0.099	0.169	0.110	0.021	—	0.099	—	0.114	0.118	0.032
yam (Dioscorea spp.) (2.1% ptn)	0.035	—	—	—	0.110	0.034	—	—	—	—	—	—
Yautia malanga (1.7% ptn)	0.023	—	—	—	0.067	0.061	—	—	—	—	—	—
Other vegetables (Ptn, N×6.25):												
asparagus:												
raw (2.2% ptn)	0.027	0.066	0.080	0.096	0.103	0.032	—	0.069	—	0.106	0.123	0.036
canned, solids and liquid (1.9% ptn)	0.023	0.057	0.069	0.083	0.089	0.027	—	0.060	—	0.092	0.106	0.031
beans, snap:												
raw (2.4% ptn)	0.033	0.091	0.109	0.139	0.126	0.035	0.024	0.057	0.050	0.115	0.101	0.045
canned, solids and liquid (1.0% ptn)	0.014	0.038	0.045	0.058	0.052	0.015	0.010	0.024	0.021	0.048	0.042	0.019
beets:												
raw (1.6% ptn)	0.014	0.034	0.051	0.055	0.086	0.006	—	0.027	—	0.049	0.028	0.022
canned, solids and liquid (0.9% ptn)	0.008	0.019	0.029	0.031	0.048	0.003	—	0.015	—	0.028	0.016	0.012
broccoli (3.3% ptn)	0.037	0.122	0.126	0.163	0.147	0.050	—	0.119	—	0.170	0.192	0.063
carrots:												
raw (1.2% ptn)	0.010	0.043	0.046	0.065	0.052	0.010	0.029	0.042	0.020	0.056	0.041	0.017
canned, solids and liquid (0.5% ptn)	0.004	0.018	0.019	0.027	0.022	0.004	0.012	0.018	0.008	0.023	0.017	0.007
cauliflower (2.4% ptn)	0.033	0.102	0.104	0.162	0.134	0.047	—	0.075	0.034	0.144	0.110	0.048
celery (1.3% ptn)	0.012	—	—	—	0.021	0.015	0.006	—	0.016	—	—	—
chayote (0.6% ptn)	0.008	—	—	—	0.038	0.001	—	—	—	—	—	—
cowpeas, yardlong, immature pod (3.4% ptn)	0.034	—	—	—	0.203	0.021	—	—	—	—	—	—
cucumbers (0.7% ptn)	0.005	0.019	0.022	0.030	0.031	0.007	—	0.016	—	0.024	0.053	0.001
oushaw (1.5% ptn)	0.014	—	—	—	0.044	0.008	—	—	—	—	—	—
eggplant (1.1% ptn)	0.010	0.038	0.056	0.068	0.030	0.006	—	0.048	—	0.065	0.037	0.019
mallow (3.7% ptn)	0.144	0.155	—	0.259	0.155	0.030	—	0.166	—	0.181	0.189	0.063
mushrooms:												

Protein Content, and Nitrogen Conversion Factor	Tryptophan (Gm.)	Threonine (Gm.)	Isoleucine (Gm.)	Leucine (Gm.)	Lysine (Gm.)	Methionine (Gm.)	Cystine (Gm.)	Phenylalanine (Gm.)	Tyrosine (Gm.)	Valine (Gm.)	Arginine (Gm.)	Histidine (Gm.)
VEGETABLES—Continued												
(Agaricus campestris)*	0.006	—	0.532	0.281	—	0.167	—	—	—	0.378	0.235	—
(Lactarius spp.)†	0.006	0.156	0.201	0.139	0.088	0.021	—	0.018	—	0.116	0.021	0.027
okra (1.8% ptn)	0.018	0.066	0.069	0.101	0.076	0.022	0.017	0.065	0.079	0.091	0.093	0.030
onions, mature (1.4% ptn)	0.021	0.022	0.021	0.037	0.064	0.013	—	0.039	0.046	0.031	0.180	0.014
peppers (1.2% ptn)	0.009	0.050	0.046									
prickly pears (1.1% ptn)	0.009	0.053	0.044	0.057	0.044	0.008	—	0.059	—	0.041	0.032	0.016
pumpkin (1.2% ptn)	0.016	0.028	0.044	0.063	0.058	0.011	—	0.032	0.016	0.045	0.043	0.019
radishes (1.2% ptn)	0.005	0.059	—	—	0.034	0.002	—	—	—	0.030	—	—
seepweed (2.6% ptn)	0.027	0.089	0.113	0.152	0.089	0.013	—	0.116	—	0.091	0.062	0.036
soybean sprouts (6.2% ptn)	—	0.159	0.225	0.265	0.211	0.045	—	0.186	—	0.225	0.225	0.133
squash, summer (0.6% ptn)	0.005	0.014	0.019	0.027	0.023	0.008	—	0.016	—	0.022	0.027	0.009
tomatoes and cherry tomatoes (1.0% ptn)	0.009	0.033	0.029	0.041	0.042	0.007	—	0.028	0.014	0.028	0.029	0.015
turnips (1.1% ptn)	—	—	—	—	0.057	0.012	—	0.020	0.029	—	—	—
waxgourd, Chinese (0.4% ptn)	0.002	—	—	—	0.009	0.003	—	—	—	—	—	—
MISCELLANEOUS FOOD ITEMS												
Vegetable patty or steak (principally wheat ptn) (15% ptn, N×5.70)	0.142	0.411	0.884	1.079	0.321	0.253	—	0.811	—	0.705	0.597	0.321
Yeast:												
baker's, compressed‡ (N×6.25)	0.122	0.655	0.655	1.151	0.914	0.248	0.120	0.607	0.580	0.840	0.536	0.353
brewer's, dried§ (N×6.25)	0.710	2.353	2.398	3.226	3.300	0.836	0.548	1.902	1.902	2.723	2.250	1.251
primary, dried:												
(Saccharomyces cerevisiae)§ (N×6.25)	0.636	2.353	2.708	3.300	3.337	0.851	0.444	1.813	2.472	2.553	1.931	1.103
(Torulopsis utilis)§ (N×6.25)	0.636	2.331	3.323	3.707	3.648	0.710	0.422	2.361	2.464	2.901	3.337	1.251

*Total nitrogen is 0.58%. This is equivalent to 2.4% protein on the basis that two-thirds of the nitrogen is protein nitrogen. If total nitrogen is used for the calculation, the protein content is 3.6%.

†Total nitrogen is 0.60%. This is equivalent to 2.9% protein on the basis that two-thirds of the nitrogen is protein nitrogen. If total nitrogen is used for the calculation, the protein content is 4.3%.

‡Total nitrogen is 2.1%. This is equivalent to 10.6% protein on the basis that four-fifths of the nitrogen is protein nitrogen. If total nitrogen is used for the calculation, the protein content is 13.1%.

§Total nitrogen is 7.4%. This is equivalent to 36.9% protein on the basis that four-fifths of the nitrogen is protein nitrogen. If total nitrogen is used for the calculation, the protein content is 46.1%.

MINERAL ELEMENTS* AND EXCESS OF ACIDITY OR ALKALINITY† PER 100 GM. OF FOODS, EDIBLE PORTION

Food	Minerals					Excess‡	
	Magnesium (Gm.)	Potassium (Gm.)	Sodium (Gm.)	Chlorine (Gm.)	Sulphur (Gm.)	Acidity (Cc. of N Acid HC1)	Alkalinity (Cc. of N Alkali NaOH)
Almonds	0.275	0.756	0.024	0.037	0.164	2.2	12.0–18.3
Apples:							
fresh	0.006	0.116	0.015	0.004	0.004	....	0.8–3.7
dried	0.029	0.557	0.072	0.019	0.019	....	....
Apricots:							
fresh	0.012	0.370	0.021	0.004	0.006	....	4.8–8.4
dried	0.062	1.024	0.100	0.021	0.031	....	31.3–41.9
Asparagus	0.015	0.200	0.008	0.047	0.051	1.0	0.8
Bananas	0.024	0.412	0.023	0.163	0.013	....	4.4–7.9
Barley, entire	0.126	0.495	0.070	0.139	0.153	6.0–17.5	....

*W. H. Peterson, J. T. Skinner, and F. M. Strong, *Elements of food biochemistry* (New York: Prentice-Hall, Inc., 1944), pp. 262-65.

†M. A. Bridges and M. R. Mattice, *Food and beverage analyses* (Philadelphia: Lea & Febiger, 1942), pp. 200-214.

‡The ranges indicated come from reports of several authors, in which variety and method of preparation influence the results in part. For individual studies and sources of data see Bridges and Mattice, *op. cit.*

Food	Minerals					Excess‡	
	Magnesium (Gm.)	Potassium (Gm.)	Sodium (Gm.)	Chlorine (Gm.)	Sulphur (Gm.)	Acidity (Cc. of N Acid HC1)	Alkalinity (Cc. of N Alkali NaOH)
Beans:							
dried	0.165	1.284	0.189	0.007	0.224	....	18.0–23.9
Lima, fresh	0.067	0.606	0.089	0.009	0.068	....	14.0
dried	0.181	1.899	0.282	0.025	0.156	....	41.6
string or green	0.032	0.288	0.012	0.045	0.024	....	4.1–5.4
Beef	0.032	0.382	0.066	0.056	0.221	8.1–38.5	....
Beets	0.027	0.235	0.040	0.040	0.017	....	8.9–11.4
Beet greens	0.097	0.390	....	....	0.035	....	....
Brains	0.016	0.269	0.160	0.155	0.130	17.7–20.7	....
Bread, white	0.034	0.110	0.517	0.602	0.083	1.5–7.1	....
Broccoli	0.024	0.352	0.030	0.076	0.126	....	3.6–4.9
Brussels sprouts	0.015	0.375	....				
Butter	0.002	0.019	§	§	0.009	0.4–4.3	....
Cabbage:	0.016	0.217	0.038	0.034	0.074	....	1.4–8.2
celery	0.011	0.400	0.028	0.023	0.013	....	....
Cantaloupe	0.016	0.243	0.048	0.048	0.016	....	7.5
Carrots	0.020	0.219	0.050	0.035	0.019	....	4.4–10.8
Cauliflower	0.023	0.292	0.048	0.038	0.074	....	1.4–5.3

§Variable. §Partly acid *in ripe.*

Celery	0.025	0.320	0.101	0.225	0.021	....	2.5–11.1
Cheese, hard	0.031	0.116	0.900	0.972	0.214	0.3–11.8	3.6–5.1
Cherries	0.012	0.125	0.015	0.004	0.018	....	1.7–7.3
Chestnuts	0.048	0.415	0.037	0.010	0.049	....	7.4–11.3
Chicken	0.047	0.402	0.054	0.034	0.303	9.6–25.4	....
Chocolate	0.082	0.400	0.019	0.009	0.114	8.1	7.9
Cocoa	0.192	0.534	0.060	0.050	0.197	....	0.7
Coconut, fresh	0.040	0.360	0.040	0.120	0.044	....	4.1–7.0
Collards	0.017	....	....	....	....	....	....
Corn:							
field, mature	0.142	0.300	0.110	0.041	0.124	....	....
sweet							
fresh	0.047	0.278	....	....	0.037	1.8	....
mature	0.121	0.415	0.148	0.050	0.146	6.0	....
Cowpeas, dried	0.265	1.305	0.036	0.019	0.250	....	....
Crabs	0.117	0.271	0.366	0.570	0.255	39.5	....
Cranberries	0.005	0.056	0.002	0.004	0.008	....	3.2
Cream	0.000	0.112	0.031	0.067	0.033	....	0.4–3.2
Cucumbers	0.020	0.170	0.026	0.028	0.011	....	3.2–31.5
Currants:							
fresh	0.031	0.208	0.015	0.010	0.021	....	0.7–8.8
dried	0.155	1.040	0.075	0.050	0.105	....	5.8–21.8
Dates	0.065	0.580	0.040	0.258	0.048	....	5.5–12.4

Food	Minerals					Excess‡	
	Magnesium (Gm.)	Potassium (Gm.)	Sodium (Gm.)	Chlorine (Gm.)	Sulphur (Gm.)	Acidity (Cc. of N Acid HC1)	Alkalinity (Cc. of N Alkali NaOH)
Eel	0.018	0.241	0.032	0.035	0.133	7.0–9.9	....
Eggplant	0.015	0.260	0.026	0.063	0.020	....	4.5
Eggs	0.009	0.149	0.111	0.100	0.233	11.1–24.5	....
Egg white	0.011	0.149	0.175	0.131	0.211	4.8–8.3	....
Egg yolk	0.013	0.110	0.078	0.067	0.214	25.3–51.8	....
Figs:							
fresh	0.020	0.205	0.043	0.037	0.017	....	....
dried	0.068	0.709	0.151	0.126	0.060	....	10.0–100.9
Fish (all kinds)	0.024	0.375	0.064	0.137	0.199	8.5–20.1	....
Flour, wheat, white	0.021	0.137	0.053	0.079	0.155	7.4–9.6	....
Frog	0.024	0.308	0.055	0.040	0.163	10.6–15.8	....
Garlic	0.008	0.130	0.009	0.004	0.318	....	....
Goose	0.031	0.406	....	....	0.326	7.7–24.5	....
Gooseberries	0.009	0.150	0.010	0.009	0.015	....	2.1–7.6
Grapefruit	0.007	0.164	0.006	0.007	0.005	....	6.4
Grapes	0.004	0.267	0.011	0.002	0.009	....	2.7–7.2
Haddock	0.017	0.334	0.099	0.241	0.255	8.5–19.7	....
Heart	0.035	0.329	0.102	0.204	0.151	27.6	....
Honey	0.004	0.051	0.006	0.015	0.003	1.1	0.4–4.6

Horse-radish	0.028	0.550	0.094	0.013	0.234	....	2.7–5.8
Kale	0.055	0.486	0.050	0.120	0.160	....	4.0–17.0
Kidney	0.019	0.240	0.238	0.376	0.148	8.4–31.0	....
Kohlrabi	0.052	0.370	0.050	0.050	0.039	....	6.0
Lamb (*see* Mutton)							
Leeks	0.037	0.380	0.036	0.110	0.056	....	5.5–11.3
Lemons	0.006	0.152	0.009	0.006	0.012	....	5.5–9.9
Lentils, dried	0.082	0.662	0.754	0.062	0.123	5.2–17.8	0.4–2.0
Lettuce	0.015	0.256	0.028	0.085	0.014	....	3.8–14.1
Liver	0.021	0.255	0.021	0.091	0.258	9.4–49.5	....
Lobster	0.022	0.258	....	....	....	38.4	....
Macaroni, dry	0.038	0.054	0.010	0.077	0.119	3.8–9.6	....
Milk:							
cow							
fresh	0.019	0.129	0.047	0.114	0.031	....	1.8–4.2
evaporated	0.038	0.258	0.094	0.228	0.067	....	4.6
powder	0.118	0.955	0.348	1.029	0.229	....	21.6
goat	....	....	0.026	0.163	....	....	....
human	0.005	0.055	....	0.058	0.142	....	....
Mushrooms	0.012	0.280	0.013	0.026	0.025	....	1.8–4.0
Mustard greens	0.016	0.330	0.020	0.090	0.142	....	....
Mutton	0.033	0.260	0.070	0.069	0.187	4.5–22.5	....
Oatmeal (rolled oats)	0.143	0.365	0.072	0.027	0.207	1.5–13.2	....

Food	Minerals					Excess‡	
	Magnesium (Gm.)	Potassium (Gm.)	Sodium (Gm.)	Chlorine (Gm.)	Sulphur (Gm.)	Acidity (Cc. of N Acid HC1)	Alkalinity (Cc. of N Alkali NaOH)
Oats, entire	0.150	0.450	0.168	0.089	0.187	....	....
Onions	0.016	0.200	0.020	0.053	0.065	....	0.2–8.4
Orange juice	0.014	0.200	0.006	0.008	0.005	....	4.5
Oranges	0.011	0.177	0.014	0.006	0.011	....	5.6–9.6
Parsnips	0.038	0.396	0.010	0.038	0.025	....	6.6–11.9
Peaches:							
fresh	0.015	0.174	0.012	0.006	0.005	....	3.8–6.1
dried	0.087	1.009	0.070	0.035	0.029	....	4.1–12.1
Peanuts	0.169	0.706	0.052	0.040	0.276	3.9–16.4	....
Pears	0.005	0.110	0.010	0.004	0.010	....	1.5–3.6
Peas:							
green	0.035	0.259	0.024	0.049	0.035	1.4–2.9	1.2–5.2
mature	0.121	0.943	0.072	0.034	0.178	0.5–3.4	1.2–10.3
Peppers:							
green	0.025	0.270	0.015	0.031	0.030	....	....
red	0.013	0.120	0.006	0.014	0.030	....	....
Persimmons	0.005	0.170	0.013	0.009	0.011	....	....
Pike	0.031	0.416	0.029	0.032	0.218	2.8–19.5	....
Pineapple	0.014	0.230	0.008	0.038	0.003	....	2.2–7.00

Plums	0.010	0.212	0.003	0.002	0.004	. . . .	4.8
Pork	0.027	0.415	0.081	0.040	0.216	7.7–28.6	. . . .
Potatoes	0.027	0.498	0.030	0.048	0.033	. . . .	7.0–12.8
Prunes, dried	0.032	0.845	0.101	0.004	0.024	. . . .	7.8–20.3
Pumpkins	0.021	0.198	0.011	0.025	0.016	. . . .	0.3–7.8
Rabbit	0.029	0.415	0.047	0.051	0.184	14.8–22.4	. . . .
Radishes	0.014	0.166	0.083	0.056	0.038	. . . .	2.9–7.2
Raisins	0.017	0.796	0.120	0.068	0.043	. . . .	23.7–27.0
Raspberries	0.018	0.141	0.007	0.010	0.012	. . . .	4.1–6.1
Rhubarb	0.015	0.392	0.010	0.070	0.008	. . . .	8.6–13.0
Rice:							
entire	0.141	0.334	0.068	0.066	0.121	. . . .	. . . .
polished	0.033	0.046	0.012	0.056	0.114	2.5–9.0	. . . .
Rutabagas	0.015	0.210	0.052	0.031	0.069	. . . .	8.5
Rye, entire	0.136	0.477	0.060	0.043	0.152	11.3	. . . .
Sardines, fresh	0.035	. . . .	. . . .	. . . .	. . . .	11.4–26.5	. . . .
Shrimps:							
dried	. . . .	. . . .	. . . .	. . . .	. . . .	. . . .	. . . .
salted	0.327	0.760	§	§	0.183	1.6	. . . .
Soybeans, mature	0.287	1.693	0.280	0.007	0.269	. . . .	. . . .
Spaghetti (*see* Macaroni)							
Spinach	0.048	0.416	0.093	0.118	0.027	. . . .	5.1–39.6

Food	Minerals					Excess‡	
	Magnesium (Gm.)	Potassium (Gm.)	Sodium (Gm.)	Chlorine (Gm.)	Sulphur (Gm.)	Acidity (Cc. of N Acid HC1)	Alkalinity (Cc. of N Alkali NaOH)
Squash	0.006	0.161	0.011	0.018	0.029	....	2.8
Strawberries	0.019	0.205	....	....	0.013	....	1.8–3.5
Sugar beets	0.041	0.440	0.130	0.180	0.021	....	3.3–9.4
Sweet potatoes	0.035	0.381	0.031	0.022	0.014	....	5.0–7.9
Tomatoes	0.016	0.277	0.013	0.048	0.017	....	5.6–13.7
Turkey	0.028	0.367	0.130	0.123	0.234	10.4–19.5	....
Turnip greens	0.079	0.300	0.260	0.390	0.051	....	2.3
Turnips	0.019	0.193	0.104	0.054	0.048	....	2.7–10.2
Veal	0.030	0.380	0.086	0.073	0.199	9.8–28.5	....
Venison	0.029	0.336	0.070	0.041	0.211	23.8	....
Walnuts	0.132	0.606	0.013	0.030	0.120	7.9–9.2	....
Watercress	0.010	0.100	0.031	0.059	0.071	....	....
Watermelon	0.006	0.071	0.012	0.006	0.005	....	1.8–2.7
Wheat:							
entire	0.163	0.409	0.106	0.088	0.175	9.7–12.0	....
bran	0.420	1.252	0.007	0.042	0.245	....	....
Yams	0.015	0.290	0.015	0.037	0.013	....	....

BIBLIOGRAPHY

Title	*Author*	*Publisher*
About Biochemistry	Esther Chapman	Thorsons Publishers Ltd.
About Food Values	Barbara Davis	Thorsons Publishers Ltd.
About Rice and Lentils	Harvey Day	Thorsons Publishers Ltd.
The Acid-Alkaline Balance	Mira Louise	New Horizons Publishers
Basic Nutrition and Cell Nutrition	R. F. Milton, Ph.D.	Provoker Press
The Complete Book of Food and Nutrition	J. I. Rodale and Staff	Rodale Books Inc.
The Complete Handbook of Nutrition	Steve and Gary Null	Robert Speller & Sons Inc.
Diet For a Small Planet	Frances Moore Lappe	Ballantine Books
Food Combining Made Easy	Herbert M. Shelton	Dr. Shelton's Health School
Handbook of Diet Therapy	Dorothea Turner	University of Chicago Press
The Hygienic Way of Life; The Argument for Vegetarianism	Hereward Carrington, Ph.D.	Health Research Series Publications
Let's Eat Right to Keep Fit	Adelle Davis	New American Library
Man and Food	Magnus Pyke	McGraw-Hill Book Company
Nerves and Muscles	Robert Galambos	Anchor Books
Nutrition and Your Mind	George Watson	
Nutrition in a Nutshell	Roger J. Williams	Dolphin Books
The Oxford Book of Plant Foods	Masefield, Wallis, Harrison, Nicholson	Oxford University Press
Proteins as Human Food	R. A. Lawrie	Avi Publishing Company
Protein, Building Blocks of Life	Bob Hoffman	York Barbell Company
Protein for Muscle-Bone-Skin	Dr. Bernard Jensen	Author

If you require further information concerning any material contained in this book (i.e. references, statistics, inter-
pretations, or biographical sources), write to the
[illegible] Institute of America, Inc., 200 West 57th St., [illegible]
York, New York 10019.

If you require further information concerning any material presented in this book (i.e. references, statistics, clinical observations, or biographical sources), write to the Nutrition Institute of America, Inc., 200 West 86th Street, #17A, New York, New York, 10024.